Estate Wines of South Africa

David Philip Cape Town & Johannesburg

Estate Wines
of South Africa
SECOND EDITION

Graham Knox

It is with appreciation that I record the assistance of Dr. J. Laszlo and Mr. D.P. Pongrácz in checking the information compiled and recorded here. In addition, the enthusiasm with which Patrick Humphreys, Nienke Esterhuizen, Annaline Moag, Tanja Humphreys, Diane Knox, Stella Bedeaux and Paul Konings went about the production of the various pieces of material is gratefully acknowledged. In the assembly of two editions of this book, the staff of the OVRI, KWV, the Bergkelder, Stellenbosch Farmers' Wineries and Gilbey's Distillers and Vintners, as well as owners and staff of each of the small producing wineries of South Africa, referred to within this volume as Estates, gave freely of their time to provide most of the information used in the compilation of this work. I thank them for their help. In addition, there have been many other colleagues who have provided some sacrifice or assistance to ensure the successful publication of this book, and their efforts are sincerely appreciated.

The concept of this book owes much to *World Atlas of Wine* by Hugh Johnson, published by Mitchell Beazley, London, and this is gratefully acknowledged.

Published by David Philip, Publisher (Pty.) Ltd., Werdmuller Centre, Claremont, Cape Province, South Africa

First edition 1976
Second edition 1982

ISBN 0 908396 02 3

© Graham Knox 1976, 1982

All rights reserved

Designed by Patrick Humphreys and Annaline Moag

The majority of the photographs taken for this edition were provided by Paul Konings, others by Andrew Pratt. Most of the photographs used from the first edition were by Maya Albrecht. Additional photographs by Gary Haselau and Roy Helmbold have been incorporated.

Jacket illustration by Annaline Moag

Official maps reproduced under Government Printer's copyright authority 5734 of 24.8.76

All maps have been reproduced from the latest editions available from the Government Printer

Typesetting and reproduction by Hirt and Carter (Pty.) Ltd., Cape Town

Printed and bound by Creda Press (Pty.) Ltd., Solan Road, Cape Town, South Africa

Contents

The South African Wine Industry   8
The Oenological and Viticultural
Research Institute   12
The KWV   16
Estate Wine   18
The Wine Seal   20
Wine Grapes   22
Cultivars/Varieties   23
The Climatic Factor   28

CONSTANTIA   29
Groot Constantia   30

DURBANVILLE   36
Diemersdal   37
Meerendal   40

LITTLE KAROO   44
Boplaas   45
Die Krans   48
Doornkraal   50

PAARL   52
Backsberg   54
Boschendal   57
De Zoete Inval   62
Fairview   64
Laborie   68
La Motte   72
Landskroon   75
L'Ormarins   79
Villiera   84
Welgemeend   86

ROBERTSON   88
Bon Courage   90
De Wetshof   92
Excelsior   97
Goedverwacht   99
Le Grand Chasseur   101
Mon Don   103
Mont Blois   105
Rietvallei   107
Van Loveren   109
Weltevrede   111
Wonderfontein   114
Zandvliet   116

STELLENBOSCH   120
Alto   122
Audacia   126
Blaauwklippen   128
Bonfoi   132
Delheim   134
Goede Hoop   138
Hazendal   140
Jacobsdal   143
Kanonkop   145
Koopmanskloof   148
Le Bonheur   150
Lievland   153
Meerlust   155
Middelvlei   159
Montagne   161
Muratie   164
Neethlingshof   166
Overgaauw   169
Rust-en-Vrede   173
Schoongezicht/Rustenberg   177
Simonsig   182
Spier   187
Uiterwyk   191
Uitkyk   193
Verdun   197
Vergenoegd   200
Vriesenhof   203
Zevenwacht   205

SWARTLAND   208
Allesverloren   209

TULBAGH   213
Montpellier   214
Theuniskraal   218
Twee Jongegezellen   220

WALKER BAY   225
Hamilton Russell Vineyards   226

WORCESTER   228
Bergsig   229
Lebensraum   231
Opstal   233

GLOSSARY   235

INDEX   238

The South African Wine Industry

Brief Outline

The history of South African wine is as old as the country. The first vineyard was planted by Jan van Riebeeck in 1652, the year he established the first settlement at Cape Town. Later, Simon van der Stel, while Governor of the Colony, was granted the land now known as Constantia, planted vineyards and made wine in sizeable quantity. Such was his capacity as a wine maker that he became the Colony's first wine merchant, buying grapes from other farmers to add to his own production. Simon van der Stel founded the town of Stellenbosch

Within 50 years of first settlement wine was being made commercially from Constantia vineyards on the southern slopes of Table Mountain.

This figure, part of Lady Anne's Bath at Groot Constantia, is believed to have been sculpted by Anton Anreith in 1791.

Less than 35° from the equator, the Cape has fewer daylight hours in summer than European vineyards.

and encouraged farmers to settle in the area and to plant vineyards.

Willem Adriaan van der Stel, Simon's son, also became Governor of the Colony and granted an extraordinary number of original title deeds to farmers established on the land in the Stellenbosch and Paarl districts.

The name of Constantia came back into prominence much later, in 1778, when the farm was purchased and re-organised by Hendrik Cloete. This man's ability as a wine maker and merchant was such that he made Constantia wine famous throughout

The summer south-east wind moderates temperatures and often condenses into cloud over mountain ranges.

Europe. It was served at the tables of kings and emperors, of whom Napoleon Bonaparte is probably the most famous. When the British imprisoned Napoleon on the island of St. Helena, one of his privileges was to be able to eat and drink in the style to which he had been accustomed while Emperor of much of Europe. He chose, for his prison table, clarets and champagnes from France, and Constantia from South Africa.

When the Huguenots fled religious

persecution in France, many thousands of them immigrated to South Africa. Naturally, there was a large number of farmers in this group, and they settled around Stellenbosch, Paarl and Franschhoek, establishing many of the nation's most famous wine farms of today. Names they brought, such as Du Toit, De Wet, Malan, Theron and Joubert, are still to be found wherever wine is made in South Africa.

The Cape's Mediterranean climate is ideally suited to the growing of wine grape varieties. Rain may be expected through the winter months from May to October, when the vines are in a state of dormancy. As it is unusual for rain to fall in the Cape between December and April, the grapes are able to ripen fully and farmers are more concerned that there should be too much sun than too little, unlike German wine farmers who face the reverse problem. The average rainfall in the Stellenbosch, Paarl, Constantia and Durbanville areas, the first districts to be settled, is approximately 65 cm per year. Many farmers in these areas today believe that this is not enough for optimum growth of vine cultivars and add supplementary water in the form of irrigation. The average rainfall in Bordeaux is approximately 90 cm per year, so there is a sound basis for this belief.

The Meditérranean climatic zone is confined to the very tip of the continent, for the mountain ranges within fifty kilometres of the sea cause the rains to fall on a relatively small area in the south and south-west parts of the Cape. Worcester, less than eighty kilometres from Cape Town, in a direct line to the north-east receives as little as 30 cm per year. This rainfall is far too little to support cultivation, and so development of wine lands to the north, north-east and east of the Drakenstein, Helderberg and Hottentots Holland ranges of mountains did not begin until dams had been built on rivers such as the Breede, and water for irrigation made available to farmers throughout the year. This did not happen until the beginning of the twentieth century.

Now the majority of South African wine farms lie in these districts, using water from pipe lines and canals to keep their vineyards alive and producing. These inland areas now supply the bulk of both table grape production and table wine production. Though vines have been grown in quantity in areas such as Robertson and Worcester since the early 1900s, Stellenbosch and Paarl based table wine producers have not considered that these areas offer serious competition in the production of high-quality wines. There have been many theories, of which the distance from the sea, the absence of sufficient, reliable winter rain and the effect of very high summer temperatures are most often used to support this attitude. But the success of table wines from inland areas in national wine shows and growing consumer acceptance of

Cape vineyards were traditionally planted in valleys. The quest for cooler temperatures has taken vines up the slopes.

The South African Wine Industry

these wines seem certain to erase this long-held belief.

Today, table wines are being produced as far north as Vredendal on the west coast and as far east as Oudtshoorn in the Little Karoo.

Dating from the first year of Jan van Riebeeck's settlement, viticulture in South Africa is over 330 years old, but unfortunately South African wine making does not have three centuries of tradition behind it. There have been many periods since 1700 when wine was not a fashionable drink or profitable to produce. During these periods, South African wine farmers became graziers, fruit growers, distillers or merchants, and ploughed under their neglected vines. There have been several long periods when wine was not produced at all; the most spectacular of these followed the destruction of virtually every vine in the land by the phylloxera insect, dating from about 1885. While still in its larva stage, the phylloxera pest's staple diet is vine roots. As this process happens underground, the first the farmer is likely to know of approaching tragedy is the death of the first vine.

Phylloxera almost wiped out many of the world's most important wine-producing vineyards, but before the last vines died, French and German wine farmers were grafting cuttings from noble cultivars on to phylloxera-resistant root stock from America. Though there were several vintages in Europe with volume severely reduced, the industry remained in business. On the other hand, South African farmers moved into different fields and there was a gap of at least twenty years before the first vines, grafted on to American root stock, were planted on a reasonable scale. Even after farmers once again began to make wine, there was little stability in the industry, and therefore no incentive to stock large vineyards until the formation of the KWV in 1918. This organisation is now so important to the present South African wine industry that a separate chapter has been devoted to it in this book.

From 1918, and the beginning of the KWV, South Africa's production of wine has steadily grown in quantity, and has also made a striking improvement in quality, especially in the production of fine table wines. Once again, South Africa is exporting wine in quantity, with sales growing year by year. The structure of the actual industry is unlike the industries of Europe and similar to those of the United States and Australia, with the important exception of the unique KWV. Wine is grown in areas that receive sufficient winter rainfall for cultivation to be successful, but also in areas that receive little rainfall. Such

Because the foothills have higher rainfall and moisture-retaining soils, many Estates are found on sloping land.

There are two Estates in the Durbanville district, with vineyards among the wheat fields, beside the expanding town.

Most South African vineyards are planted on alluvial soils in low-rainfall areas. Farm dams provide additional water in this coastal valley.

A patch of the famed Durbanville red soil is being prepared in summer for planting with vines in late winter.

Many South African vineyards are grown on low-potential soils. When vines are grown without trellising, the smaller crop can have high quality.

irrigation areas are important producers in California and Australia also.

In South Africa wine farms fall into three separate categories.

Some farms have the necessary capital, and produce grapes of sufficiently high quality, to operate a modern, fully equipped cellar, where all their wine is made. First among these are the Estate farms, which are the subject of this book. The others sell wine made in the farm cellar to wine merchants.

Many other farms belong to district co-operatives, of which there are no fewer than five registered in the Stellenbosch area alone. A member of such a co-operative supplies his newly harvested grapes to the co-operative cellar, where it is made into table wine, sweet wine, rebate wine or distilling wine. Most of this wine is later sold by the co-operative to wine merchants and distillers. A much smaller amount of wine is sold to the retail trade for sale to the public under the co-operative's name.

The third group of farms have their wine made by a wine merchant, who buys the produce at grape stage. Each of these merchants has a large, modern, well-equipped and efficient cellar, where the grapes are made into table wine, sweet wine, rebate wine and distilling wine, depending on their quality and type. Another important side of a wine merchant's activity is the purchasing and marketing of most of the huge quantity of wine made by co-operatives, including the KWV. There are two major wine merchants operating in South Africa today, the Stellenbosch Farmers' Wineries with its associate companies, and the Oude Meester Group. Several smaller merchants include Gilbey's, Douglas Green, and Union Wine.

In a major re-organisation of financial interests in the South African liquor industry in 1979 the Stellenbosch Farmers' Wineries and the Oude Meester Group were brought under the control of Cape Wine and Distillers Limited, a body jointly controlled by the KWV and Rembrandt. At this time, it was agreed by the major liquor merchant houses, with the exception of Union Wine, to sell off the majority of the retail bottle stores under their control and allow more open trading at consumer level. This brings almost 80 per cent of the wine produced in South Africa under one controlling body, but allows the smaller merchants and private Estates less restricted access to the market.

Most Estates have between one and two hundred hectares under vineyard. Few of them are able to sell the quantity of wine that this allows them to produce, and the greater part of their annual crop is sold in bulk to one or more of the merchants.

The South African Wine Industry

The Oenological and Viticultural Research Institute

The OVRI is situated on the farm Nietvoorbij, in good wine country, just to the north of Stellenbosch.

Flowering is probably the most sensitive stress stage in vine development, and disease control monitors susceptibility.

An important test programme has over 80 different strains of rootstock. The grafts may be seen just above ground level.

Behind every expanding industry there must be continual development and improvement of the systems, materials and knowledge used. The South African wine industry is growing at such a remarkable rate that research to aid its development is hard-pressed to meet the continual demand for new knowledge.

South Africa has an up-to-date, fully equipped oenological and viticultural research centre. Research is basically divided into two sections. Oenology is the science of wine making (or cellar technique), and viticulture is the science of vine growing. But even these two categories must be split into many more to be able to describe the aims and purpose of the Institute. The viticultural aspect of the organisation is concerned with finding the best uses of soil, climate, vineyard preparation, vine root material, fertiliser, quality-wine-producing vine cultivars, irrigation, and pest and disease control. The oenological section conducts research into the most effective technological cellar practices and the applications of wine chemistry and microbiology in preparing various types of wine.

All the resultant information must then be relayed economically and effectively to the wine farmers, who have neither the finance nor the time necessary to do research on the scale carried out by the Institute.

Though there are approximately 10 000 wine farmers in South Africa, the results of the Institute's work in established wine-farming areas can best be seen on Estate farms, where an improvement in quality in any aspect of wine production first becomes evident. An Estate farmer implementing any improvement on his farm is likely to be the first to see a financial reward for his efforts.

For instance, the Estate farmer has special needs with regard to plant cultivars. He has to have something specially suited to his circumstances, his soils, climate and situation. He has to produce wines with a character typical of his particular Estate. He has to make recognisable wines. Thus the Estate farmer may be interested in cultivars for which there is no great demand from the industry as a whole. (Chardonnay, with its striking flavour and aroma but limited production, is such a variety.)

The Institute tests a cultivar in two ways. The first is to evaluate its general qualities on the Institute's own farm, Nietvoorbij, a 200-hectare wine farm of great value, situated just outside the town boundary of Stellenbosch. The second is to find the soil, climatic and environmental conditions under which the vine will produce its best wine. To achieve this latter aim, the Institute has experimental vineyards at Robertson, Vredendal, Vaalhaarts, Upington and Oudtshoorn, as well as on a number of farms in different districts. At Nietvoorbij, the Institute maintains a large number of particularly well-equipped laboratories, and an experimental winery which is one of the most modern in the world for small-scale wine production.

The Institute has a programme of classifying the many dozens of different soil types found in the winelands, which include the northern areas along the Olifants River, the inland areas of the Orange River and obviously also the coastal lands of Stellenbosch and Constantia. This particular test programme is closely linked with research into better cultivars. To give some idea of the size of the entire soil project, the Stellenbosch area alone has approximately fifty recognisably different soil types.

The control and elimination of vine and grape disease are among the Institute's most important projects. The vine species known as *Vitis vinifera*, which produces the finest wine, is especially prone to disease and destruction by pests, such as phylloxera. However, there are many other species of *vitis* which are resistant to phylloxera, but which produce grapes of inferior quality. Today, almost every wine-producing vine in South Africa is a combination of a phylloxera-resistant rootstock and a vine of the *Vitis vinifera* group, grafted together.

A test programme has been under way for some time to find and develop better rootstocks. The programme is at present working with over 80 different strains of phylloxera-resistant rootstocks. This test programme, while being mainly directed against nematodes, fungus, phylloxera and other root diseases, is also linked with the cultivar project, as some rootstock cultivars combine better with one scion cultivar than another.

A further programme under inves-

The varieties and cultivation best suited to different soil types are part of the OVRI's test programme.

Cuttings from selected vines in producing vineyards are prepared for grafting during the early winter months.

The leaf and stem growth from the grafted vine has originated from a bud at the top of the planted twig.

The stems of the roostock and the scion material are grafted together and then bound to ensure union.

This Metallica rootstock cutting is planted, ready for vineyard grafting, as an alternate method to nursery grafting.

The Research Institute

Tests to determine the optimum affinity between different varieties of rootstock carried out in all areas.

The Institute has bred many new scion cultivars, the most promising of which are being evaluated.

The effects of irrigation on the physiology of the vine and the quality of the grape juice are being evaluated.

tigation is an evaluation of the different types of trellising, to find which type suits which cultivars best, under specific climatic conditions. As the quality of hand labour on South African wine farms decreases and becomes more expensive, mechanical harvesting of grapes is being tested under local conditions. This machinery has been developed mainly in the United States, but its use there has been restricted by labour union problems. Mechanical harvesting has proved to be a success in the Argentine, a very large producer of wine, and Australia, where wine farming labour has always been scarce.

The Institute's work on vineyard irrigation has been of great use to the industry, which has moved from flood and sprinkler irrigation to large-scale implementation of measured drip and micro-spray irrigation. Research programmes are measuring the effect of irrigation on the micro-climate of the vine and the quality of the wine it produces.

In the field of wine making, methods of extracting colour and flavour from the skins of red cultivars constitute another test programme. The Institute is evaluating the effectiveness of some of the methods currently in use. These include manual 'punching' of the cap of skins that form on the top of fermenting must, down through the top layer of the liquid. Another involves pumping the must from under the skins back over the top and down through the cap. A further method involves the retention of the cap below the surface of the liquid by using strongly anchored, high-strength, fine-mesh pilchard nets. A more sophisticated method uses high-pressure sealed tanks, where the natural build-up of CO_2 under the cap is used to break up the cap and mix it with the liquid at controlled intervals. Yet another involves the use of rapid heating of the juice and cap, followed by an equally fast cooling to normal fermentation temperature.

Tests are run on the affinities of cultivars for different soils as well as vine root and water supply relationships.

Because many new vineyards are on high slopes, the influence of micro climates is being tested.

To find the optimum time for harvesting a grape to make wine, the Institute's work in the field of sugar/acid ratios and sugar/pH ratios has proved extremely successful.

Other programmes include an attempt to control diseases that affect rootstock, such as fungus that strangles some varieties of rootstock under wet conditions. The Institute is testing the many different methods of soil preparation popular in this country. It is examining the influence of the size of each vine's crop and its effect on wine quality, and in another programme the effectiveness of the use of plastic strips along rows of young vines. The plastic keeps the temperature at root level relatively constant and reduces the amount of evaporation in the immediate vicinity of the young roots.

Further programmes are testing the influence of temperature on wine quality, the influence of long-distance road transport of grapes on wine quality, the influence of climate and even micro-climate on wine quality, the use of enzymes in red wine production to extract greater colour, the most efficient way of 'cleaning' must, and the effect of the different 'cleaning' methods on wine quality. For example the cold settling process is being compared with the use of a centrifuge. One of the Institute's largest test programmes covers a wide range of disease and pest controls. Diseases and insects that affect the vine, the roots of the vine, its leaves and its grapes are under careful examination in an attempt to control the great number of casualties that occur throughout the industry, in different areas, under different conditions and in different vintages. Disease control is one of the most important functions of all research undertaken by the Institute, as disease is one of the greatest factors inhibiting optimum crop production. Major new research projects involve study of the physiology of the vine, vine root and water supply relationships, flavour substances in wine, soil influence on wine quality and the influence of micro-climate on grape composition and wine quality.

The Research Institute

The KWV

KWV technical staff, July 1938: (from left) Messrs. J. Goossens, J. de Villiers, C. Snoek, D. de Wet Theron, Dr. A. I. Perold, Mr. F. W. Myburgh, Dr. C. Niehaus.

A detail from the pediment above the entrance to the KWV, Paarl.

In 1915, 1916 and 1917 South African soldiers were dying in the fields of wartorn Europe and South African wine farmers were pouring wine down drains. They were unable to sell it at any price. There was a severe depression in the industry, with the possibility that it, too, might die in the war years. The problem was severe overproduction, for farmers were producing far more than the then tiny South African domestic market was capable of consuming. The wine merchants of the day were able to dictate prices, and consequently farmers competed with one another to see who could sell most of his crop and keep in operation.

Charles Kohler, a retired dentist who owned a wine farm in the Paarl Valley, persuaded sufficient farmers to group together to make an impact on the standard practice of wine purchase, and in 1918 they formed South Africa's first wine farmers' co-operative. They named their organisation the Ko-öperatiewe Wijnbouwers Vereniging van Zuid-Afrika Beperkt, which has become far better known as the KWV. Surprisingly, Kohler was able to persuade almost all wine farmers to speak with one tongue, and the fledgling operation kept in business many wine farms that were on the point of bankruptcy.

Though the formation of the KWV was a great achievement, even greater problems lay ahead in deciding how it would actively help individual wine farmers and the industry as a whole. As part of its constitution, the KWV agreed to take all surplus production at a price of half the going rate for good wine. Its constitution restricted

Dr. C. W. H. Kohler, founder and first Chairman, KWV, 1918-52.

the KWV's future sales to export markets north of the Equator.

During those first years the KWV had little money, few staff, and restricted facilities, and processed all the surplus produce it bought into distilled ethyl alcohol. On the other side of the country, during the last two years of the war, Natal sugar farmers were distilling ethyl alcohol from unwanted molasses. Together, the wine and sugar industries produced a combustible spirit which was used as a substitute for almost unobtainable petrol. This product, named 'Natalite', was the KWV's first venture. It was, unfortunately, a commercial failure, being far too expensive to compete with petrol once war blockades had been lifted. The KWV however went on distilling wine and every tank and sealable container that they could obtain was filled to capacity. Much of this spirit was to remain on the premises until the Second World War, when it was sold as industrial alcohol.

In 1924 the KWV began its second project in the disposal of excess wine production and began to produce and mature brandy. This venture was far more profitable than the first, and the KWV has become one of the largest producers of brandy in the world, though spectacular results did not show for decades. The third project was the production of table wine to be sold in export markets. During the first twenty years of its existence the KWV developed a range of products for export sale that included sweet and dry fortified wines, dry table wines, brandies and liqueurs.

In 1937 Dr. Charles Niehaus joined the KWV and was given the task of improving the quality of South African sherry and encouraging its production. Within a few years the KWV was producing sherries that rivalled the best products of Spain, and exporting millions of gallons of sherry annually. After the success of this project, Dr. Niehaus was appointed assistant wine expert to the famous Dr. Perold, and later when he became chief wine expert, turned his attention to improving the quality of table wine for sale to export markets.

16   The South African Wine Industry

A view of the headquarters of the KWV in Paarl.

The KWV was at that time making two types of dry red table wine for export, which were sold under the names Red South African Table Wine, light-bodied and full-bodied. The light-bodied wine was a blend of the product of Cabernet Sauvignon and Cinsaut cultivars. The full-bodied wine contained a blend of wines made from Pontac, Cinsaut and several other varieties, some of which are no longer grown. Dr. Niehaus paid particular attention to the full-bodied wine, encouraging farmers to plant Shiraz in greater quantities, and substituting it for Pontac, which had been the base for the old blend. Later the wines were given more commercial names and sold as Roodeberg No. 1 and Roodeberg No. 2.

Eventually the trade name Paarl was adopted for all KWV table wines, with the light-bodied red wine be-

Dr. Charles Niehaus, founder of the South African sherry industry.

coming Paarl Cabernet and the full-bodied wine becoming Paarl Roodeberg. This latter product was to become the most popular South African wine in those countries to which KWV sold wines in bottle. (Most KWV wine sold to Great Britain is shipped in bulk.)

The KWV's role in the life of today's South African wine farmer can here be told only in simplified terms. The organisation operates a production quota system, applying to all farmers, each of whom has a maximum production quantity. Those farms not directly owned by wine merchants are paid by the KWV for produce that they deliver to merchants. This roundabout system is fundamentally the same as that developed in the 1920s. With payment passing through the KWV, and all records of production, including size of farm, area of vineyards and number of vines of each variety, being submitted by law to the KWV, the organisation is now a colossus astride the wine industry.

In 1979 the KWV, together with the Rembrandt organisation, engineered a take-over of the South African Breweries wine interests and formed a body that jointly controls the Stellenbosch Farmers' Wineries and the Oude Meester Group. (The KWV is a statutory body, being the subject of an Act of Parliament in 1925.)

The KWV

Estate Wine

Estates are geared to the production of small quantities of specialised products. Several have commenced production of bottle-fermented sparkling wine.

Top quality dry wines can be made only from healthy grapes. All Estates have vigorous crop protection programmes.

In a good growing season, excess leaves and bunches have to be trimmed back to allow remaining grapes to ripen fully.

The Estate farms are the prime wine farms of the country. They make their own wine to demonstrate that it is of superior quality and may fetch a high price. Not all wine made on an Estate is sold as Estate wine. Wine merchants purchase the surplus produce of farms such as these for blending with inferior wines. Not all Estate wine is fine wine, nor is it all necessarily superior to the produce of the wine merchants' cellars, but the best wines of the country are grown and made on the Estate principle.

The most exact definition of an Estate wine is that it must have been made on the farm on which it was grown. The man in charge of the vineyards and grape production may not necessarily be the wine master in the cellar, but the two activities must take place on the same property. The produce of that farm may not leave the boundaries of that farm until it has been made into finished wine, if it is to be certified as an Estate wine. The classification and certification systems employed by the Wine and Spirit Board under the 'Wines of Origin' legislation apply as strictly to Estates as they do to any other wine producer. If the wine has indeed been grown and made on the defined Estate property, the Estate is entitled to use the word 'Estate' at the very top of the wine seal on every bottle of Estate wine. The controls necessary to implement this certification are detailed in the next chapter. We are here concerned with what an Estate farm has to be to live up to its name.

To qualify for use in the production of Estate wine, each grape used in the wine must be grown within the classified boundaries of the Estate farm. No grapes may be brought on to the farm for use in the production of Estate wine. This control serves several purposes. The 'Wines of Origin' laws have the broad purpose of separating wines into areas of origin, and keeping them there until they are finished wines ready for bottling.

Climate and locality have as great an effect on wine type and style as cellar technique has; for example, a Riesling from Tulbagh is markedly different from a Riesling from Stellenbosch. An Estate is simply an even smaller area within the legislated 'origin' areas. To illustrate the importance of locale and cellar technique, a Riesling from one Tulbagh Estate will differ from a Riesling from another Tulbagh Estate, firstly because each farm has a different soil and microclimate, and secondly because it is in the Estate farmer's interest to make his wine noticeably different and individual when it is compared with even his neighbour's wine. It is in this respect that a flourishing Estate wine industry is to the advantage of the wine industry as a whole.

The progressive Estate farmer, the only type who will stay in the Estate business, will be continually searching for better cultivars, or cultivars that are more suitable to his location and soil types. He will also be aware of new developments in the improvement of wine quality in the areas of soil cultivation, vineyard preparation, methods of harvesting, cellar equipment, cellar technique and wine maturation, and will make faster use of this knowledge. There are many methods by which an Estate farmer can improve his crop and his product, and he is more likely than most other wine farmers to search them out and implement in a short time those he considers practical and worthwhile. The Estate farmer is the first to see the results of his improvements in the form of better wine and the resultant cash return. By being in charge of a relatively small area of vineyard, he is theoretically able to supervise more fully any developments, changes or improvements that he implements. He is able to make earlier use of research into improved wine quality. He has good reason for doing so, as the results may be seen by the public in his wines, bearing his label.

There are no limits on the types of wines an Estate producer may make, beyond the source of his grapes. There is a minimum standard of quality that all Estate wines must meet, in analysis and taste, before they are certified by the Wine and Spirit Board, but, considering the competitiveness of the Estate wine market, no Estate wine farmer would be likely to attempt to sell any wine that did not reach such a quality level. An Estate wine farmer may make fortified wines, unfortified wines, sweet wines, dry wines and all shades in between. His market is his only control in these areas. If there is a demand at a worthwhile price for a particular type of wine and he has the ability to produce it, he will do so.

South African Estate farms produce sweet fortified wines, dry and semi-sweet wines, pure cultivar and blended cultivar wines, succulent 'noble rot' wines, sparkling wines made by the traditional 'Champagne' method, port and sherry wines and many others.

Though an Estate wine must be made on the Estate from grapes grown on the Estate, it may be bottled elsewhere, if supervision of its movements at all times prior to bottling and records of its history, from vineyard to retailer, are available to the Wine and Spirit Board.

There is no real limit in area to the size of an Estate farm, which may be so small as to contain only one vineyard and a small cellar, or may be a conglomerate of several farms, with adjoining boundaries, that have been made into one farm. The registration and operation of this latter type of Estate are under the strict control of the Wine and Spirit Board. It must be shown that the proposed Estate, has relatively uniform climate and environment. All of the ground must lie within one 'Wine of Origin' area.

When the laws governing Estates were passed by an Act of Parliament, several small Estates were at that time in operation selling wine as Estate wine that had been produced on two or more properties that did not have adjoining boundaries but were owned by the same person. In these several special cases, involving farms owned by the one person and operated as one unit before 1 January 1973, the law made exceptions regarding the rules involving adjoining boundaries.

The registration of wine Estates has been a legal requirement of the Wine and Spirit Board since 1973. Only those properties formally registered may use the term 'Estate' on their labelling and in promotional publicity for their wine. However, there are several farm cellars, producing limited quantities of high-quality wine from grapes grown under their control, who are unable to meet the specifications for a single Estate registration and who trade as unregistered Estates. We have included a number of these in this volume and describe them as such in the text.

The first requirement of an Estate is limited production and personal supervision. One small cellar uses only grapes from this vineyard.

Estate Wine

The Wine Seal

That familiar little sticker at the top of most South African wine bottles is solely a guarantee that the wine in the bottle has been certified by the Wine and Spirit Board. The certification process begins as soon as the grapes begin to ripen. To have the wine certified, the Wine and Spirit Board must be informed of the quantity and variety of grapes that will be harvested, and their place of origin. When a producer wants to state on his label that his wine is from a particular area, is from a specific vintage or has been made from a particular cultivar, such as Cabernet Sauvignon or Riesling, the process of certification begins.

Information required by the Board from the producer (who must maintain continually up-to-date records) includes: an estimate of production in tonnes of grapes, well before harvesting; the exact quantity of wine made from these grapes; and information regarding the way in which the wine will be used (e.g. to be sold as a cultivar wine, or as a wine from a particular vintage, or as a wine of origin, or a combination of all three) together with samples of the wines. Once this information is received, the wine is evaluated by a tasting panel. The wine is then provisionally approved and seals are provided for labelling.

The number of wine seals granted depends on the number of seals requested, plus figures previously submitted, such as the bulk amounts of wines made, and the number of bottles made from them, and naturally all the figures must tally. Certification begins as soon as the grapes begin to ripen and does not finish until the bottled wine is ready for marketing.

The Wine and Spirit Board makes two assessments on a quality basis. All wines are assessed on their basic merit and outstanding wines may be judged 'Superior'. However, beyond this point certification is only involved with the relationship between the contents of the bottle and the label on the bottle. For instance a wine that is not produced on a registered Estate for marketing as Estate wine may not use the word 'Estate' anywhere in the labelling, packaging or promotion.

CULTIVAR

Wine may only be sold under a cultivar if the wine in the bottle contains at least 75 per cent of wine produced from that cultivar. At this date, there are still some exceptions.

Wines may be labelled Cabernet Sauvignon, Cabernet Franc, Pinotage, Cape Riesling, Shiraz or Sylvaner if the bottle contains 50 per cent of wine made from those cultivars. From 1 January 1984 these cultivars will also have to comply with the 75 per cent rule.

VINTAGE

At least 75 per cent of the wine in a bottle carrying this stripe was produced in the vintage year stated on the label. Most vintage wines contain only wine made in a particular year, but an allowance has been made for a producer who has a quantity of wine left over from one year's bottling, which is of insufficient quantity to be economically bottled by itself as a vintage wine of that year. Because of the 75 per cent content law, he can blend this wine with his next vintage, without affecting the distinctive character of that vintage.

This wine will contain blends of wines from more than one vintage and more than one cultivar.

This wine will be a blend of more than one cultivar and the produce of the vintage on the label.

The South African Wine Industry

ORIGIN

South Africa's wine country has been zoned into five regions: Boberg, Breede River Valley, Coastal Region, Klein Karoo and Olifants River. These regions have been further divided to provide Wine of Origin districts: Constantia, Douglas, Durbanville, Overberg, Paarl, Piketberg, Robertson, Swartland, Stellenbosch, Swellendam, Tulbagh and Worcester. Certain areas have been further subdivided, for greater definition of locality, into wards and those declared up to date of publication are: Aan-de Doorns, Andalusia, Benede-Oranje, Boesmansrivier, Bonnievale, Cedarberg, Eilandia, Franschhoek, Goree, Goudini, Groenekloof, Hoopsrivier, Le Chasseur, McGregor, Nuy, Riebeekberg, Riverside, Scherpenheuvel, Simonsberg-Stellenbosch, Slanghoek, Vinkrivier, Walker Bay.

Vintage
Origin
Superior
Estate Wine
Cultivar

This wine will be made from the cultivar stated on the label and be a blend of several vintages.

ESTATE

Only registered Estate farms qualify for this addition to the seal. The laws governing the composition and function of Estate farms are contained in the previous chapter. As every Estate lies totally within a wine-growing 'area', i.e. no Estate lies across the boundary between one 'area' and another, all Estates automatically qualify for the 'origin' classification. Estate farmers must gain other certification in regard to cultivar and vintage in the same way as other producers.

The seal guarantees the accuracy of the details on the label, in this case:

SUPERIOR

Though wine submitted for certification must be of a certain minimum standard to be issued with wine seals, the Wine and Spirit Board does not classify wines on ratings of merit. Nevertheless, when the certification programme was drawn up for legislation, a decision was made to provide an extra incentive for producers to make outstanding wines. The most effective incentive, naturally, is a higher price, and the best method of selling a wine at premium price is to give the wine a reliable, responsible award that can be shown on the seal and on the label. It is imperative that no wine of anything less than truly outstanding quality be granted the award 'Superior', so the panel of tasters who have the authority to bestow this award are highly skilled, practised and experienced individuals involved in the wine industry. As a precaution against any prejudice, they are only told the cultivar and vintage specified on the application, and are not told the specific source of the wine until after they have made their judgement.

The 'Superior' classification is linked to wines of origin. A wine may be a blend of different vintages and of different cultivars, but it must have been grown and made within one area to be certified 'Superior', thus the official title of such a wine is 'Wine of Origin Superior', often abbreviated to 'WOS'.

The Wine Seal

Wine Grapes

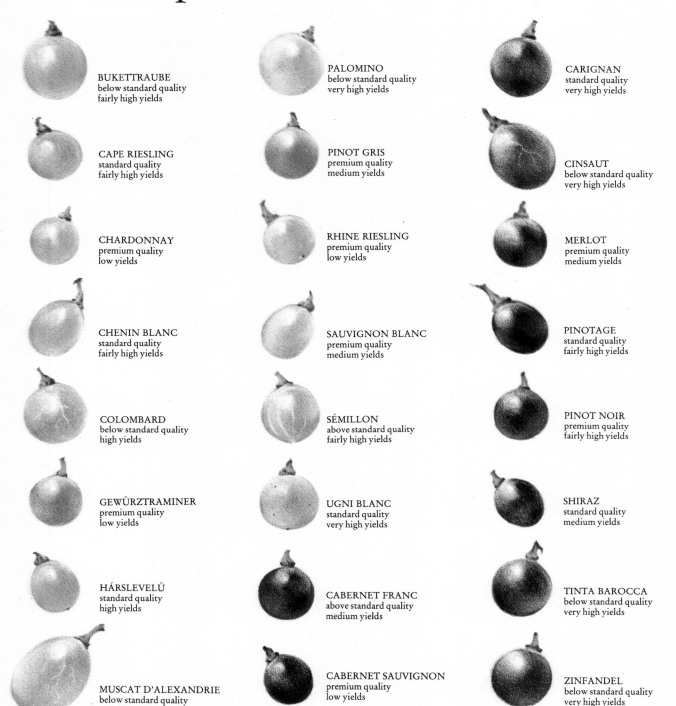

Wine grapes are almost always smaller than table grapes. They are usually juicier than table grapes, with a strong varietal flavour. The premium cultivars usually have small berries, with thick, tough skins. When ripe, wine grapes are much sweeter than table grapes, which makes them less pleasant to eat.

To be capable of forming sufficient alcohol to make wine, the grape should be ripened until at least 18 per cent of its content is sugar. If a late harvest style wine is desired, sweeter grapes are required to give extra body to the wine and to ensure that sufficient alcohol is produced.

Cultivars/Varieties

A cultivar is a cultivated variety of grape vine. To be classified as a separate cultivar the variety must produce a wine that regularly displays its own varietal characteristics. In other words, the wine made from one cultivar must be recognisable as the product of that cultivar. A Riesling must have the typical aroma and flavour of a Riesling, wherever it is grown and made. Cultivars, being the grapes themselves, are the essence of a wine industry, but their names are possibly even more important to the South African consumer than they are to the wine consumers of most other countries. Most countries sell their wine as being of a type particular to an area. This is occasionally achieved by using the product of one cultivar (for instance, most good champagnes are made from the Pinot noir variety) and more often by blending the wine made from two, three and often more cultivars.

Certain countries, such as Germany, Australia, and the U.S.A., use cultivar names on their labels. The name of a noble cultivar, such as Cabernet Sauvignon, has so much influence on the purchasing habits of the South African wine consumer that legislation has been passed to ensure that a minimum content of wine from the cultivar stated on the label is in the bottle. More information on this subject will be found in the chapter 'The Wine Seal'.

In Europe the quality of the wine is generally indicated on the label by its origin, the area, district, or even the individual farm or estate or château from which it comes. Most of South Africa's best wines are known by cultivar names. To know South African wines, one must know the vines from which they come, and how these differ from one another.

WHITE VARIETIES

BUKETTRAUBE

Of unknown origin, this recently introduced hybrid has been fairly extensively planted in South Africa, though it is not grown in Germany, where it originated. It grows well when pruned short, bears well and ripens earlier than Steen, and has good sugar and acid content. Though the quality of its wine is reasonably good, it is not considered to be in the class of Rhine Riesling, Chardonnay or Sauvignon blanc. Its main attribute is the outstanding aromatic bouquet of its wine.

CAPE RIESLING

To distinguish it from more illustrious Rhine Riesling, to which it is not related, this cultivar is also known as South African Riesling. Though it has made some of the Cape's best white table wines for several decades, it falls now in the shadow of the classic varieties. The wines often tend to be neutral, with insufficient varietal character. The vines are susceptible to fungus diseases and the quality of the wine is distinctly affected by overcropping and sustained irrigation.

CHARDONNAY

In Europe, Chardonnay is most famous for its exclusive use in two of France's most famous white Burgundies, Montrachet and Chablis. Chardonnay is also grown in other districts of Burgundy and is the second grape of Champagne.

Although most champagne is made from Pinot noir, Chardonnay is the only variety used in the famous 'blanc de blancs' champagne. Chardonnay has never been very popular in South Africa, because locally available material was very susceptible to millerandage. It ripens its grapes early, with high sugar content and high natural acidity, and produces a fine quality wine that improves greatly with maturation. It adapts well to warm climatic conditions, as has been proved in California, and produces there, as in France, the noblest white wines in the world.

CHENIN BLANC, STEEN

Chenin blanc is one of the oldest varieties grown at the Cape. It became the cornerstone of the industry during the present century, and today provides nearly a quarter of all wine made in South Africa. This is unfortunate, as no variety is sufficiently adaptable to be able to provide quality wines and bulk wines to the complete satisfaction of both markets. Chenin blanc was a safe bet during the growth of the lower priced market and is severely overplanted. It is an extremely versatile variety but is not capable of producing really high quality dry wines. It is ideally suited to the production of semi-sweet and sweet table wines. To produce noble wines, Chenin blanc vines should be restricted to medium production and the grape must should be prevented from fermenting dry.

COLOMBARD, COLOMBAR

In France, this cultivar is second only to St. Émilion as a grape for brandy production. Imported to South Africa for this purpose, Colombard has been found to make medium quality table wines in certain areas, with a high level of production. It has a naturally high acidity and often produces unexpected flavour.

GEWÜRZTRAMINER

A clonal variation of Roter Traminer, providing greater varietal character in the aroma and flavour of its wines. Originally from the Rheinpfalz, it has proved especially popular with the grape growers of Alsace, producing noble wines with fairly high alcohol and some residual sugar. Early plantings in the Tulbagh district used Roter Traminer and produced rather neutral wines. More recently planted vineyards have Gewürztraminer vines originating in Alsace, Germany and California and have proved to be very much more distinctive. Gewürztraminer needs to be fully ripe to make full-flavoured wine and, when grown in warmer areas, loses acidity quickly when approaching the desired sugar content.

HÁRSLEVELÜ

Of Hungarian origin, this recently imported variety has features that will appeal to both grower and wine lover. It grows vigorously in all soils and adapts well to a dry climate. The grapes ripen with plenty of acidity and the wines are full-bodied with distinct varietal character.

KERNER

A cross between Rhine Riesling and Trollinger, Kerner is grown in Germany because it ripens earlier and is more productive than Rhine Riesling. Introduced to South Africa during the early '70s because of its illustrious parent and the high acidity of its grapes, the Kerner vineyards on several Estates have so far failed to produce significant wines. The wine is hard, neutral and acid while young, and requires a degree of sweetness and some age for best effect. However, Kerner has suited the prevailing preference for 'simple' white wines.

MÜLLER-THURGAU

The initial popularity of this variety in Germany was probably caused by its release soon after the unexpected early death of its breeder, Dr. H. Müller. An artificial hybrid, produced by crossing Rhine Riesling and Sylvaner, Müller-Thurgau was selected from numerous seedlings. It is widely cultivated in Germany because it ripens its grapes before any other variety and does not rot like Rhine Riesling. It produces grapes with moderate sugar content and low acidity—and does not appear suitable for the warm climate of South Africa.

MUSCAT D'ALEXANDRIE, HANEPOOT

Capable of producing high volumes of low quality wine, this table grape variety can make a powerfully flavoured and highly aromatic wine in the muscat style, if grown and harvested prudently.

Small quantities are often used by table wine producers to blend with wines lacking in character. Hanepoot's greatest values are the production of table grapes and richly flavoured fortified wine.

MUSCAT OTTONEL

In South Africa, this noble variety can bear more grapes than it is able to support and ripen. It has fairly large berries, produced in small bunches, and can attain high sugars with low natural acidity. It is a very early ripening variety and produces fine quality wines, with a distinctive muscat aroma, which are normally used in Europe for blending.

PALOMINO

PALOMINO

A universally popular cultivar, Palomino is grown throughout the world, under many different names. In Europe, Palomino is responsible for Spain's great sherries. Though Palomino was intended to promote the development of a South African sherry-type wine industry, it is now used mostly in the production of *vin ordinaire*, and wine for quality brandy. Curiously, its local names are Frans and White French, though France is one of the few countries that grows very little of it. In South Africa, Palomino grows well, bears well and is resistant to virtually every calamity that can happen to a vine. However, its grapes have a low acid and sugar content, and there is no real place for it on a wine Estate.

PINOT GRIS, RULÄNDER

A colour mutation of Pinot noir, this noble variety produces white wine with a golden yellow colour. It is capable of producing full-bodied wines with rather low acidity and is sufficiently flexible to make excellent dry wines as well as late harvests.

RHINE RIESLING, WEISSER RIESLING

One of the most famous white cultivars in the world, this true Riesling is responsible for most of the finest wines of Germany. As with most fine and delicate things, the Rhine Riesling cultivar is vulnerable. It is susceptible to *Botrytis*, is heavily affected by sunburn in the hot summer, and it has small bunches that are difficult to harvest. However, being the origin of mildly aromatic and distinctively flavoured white wines, this cultivar is very important to the Estate wine industry. Most Estates producing quality white wine have young Rhine Riesling (also known as Weisser Riesling) vineyards in production, and the cultivar is destined to become one of the most important in the country.

SAUVIGNON BLANC

This is one of the two most important white varieties in the Graves and Sauternes districts of Bordeaux. Together with Sémillon, Sauvignon blanc produces the classic sweet wines of Sauternes. Elsewhere, Sauvignon blanc is the premium variety grown in the upper Loire Valley, where it has become famous as the source of Sancerre and Pouilly-Fumé wines. The variety is becoming increasingly popular throughout the vine growing world, producing excellent, characteristically aromatic, dry white wines, often with a 'grassy' flavour. It has an ability to grow and produce fine wines with acidity and flavour under warm and dry conditions.

SÉMILLON

Most famous for its use with Sauvignon blanc in the production of the classic sweet wines of the Bordeaux region, Sémillon is an under-rated variety in South Africa. It is capable of making high quality wine in cooler areas, if grown prudently, without irrigation. Sémillon is used to add body and character to more neutral wines, like Cape Riesling.

UGNI BLANC, ST. ÉMILION

Known in Italy as Trebbiano, this most widely planted variety in the world has shown that in South Africa it can produce large volumes of medium quality table wine. The wines have a naturally harmonious blend of alcohol, flavour and acidity that is only imbalanced by encouraging these generous vines to overbear under excess irrigation, producing low-alcohol, flat wines. It is grown almost exclusively under irrigation in inland

Cultivars/Varieties

areas and produces wines for the brandy and medium-priced blended table wine industries.

WELSCH RIESLING, ITALIAN RIESLING

Source of some of the best white wines of Eastern and Central Europe, this versatile variety has shown that it can adapt to warmer climates. The young wines are delicate, well balanced and have good ageing potential.

RED VARIETIES

CABERNET FRANC

This locally little-known brother of Cabernet Sauvignon has been grown as isolated individual plants within Cabernet Sauvignon vineyards in the Cape for a long time, but separate vineyards of Cabernet franc are a recent development. There is a pronounced difference between the two cultivars. Cabernet franc is much more fertile, ripens a little earlier and produces wine that is thinner and matures more quickly than Cabernet Sauvignon. The wine is nevertheless of good quality, and blends well with Cabernet Sauvignon.

CABERNET SAUVIGNON

Probably the shiest bearer amongst all wine varieties in production in South Africa, Cabernet Sauvignon owes its popularity to the noble red wine that it produces. Cabernet Sauvignon makes an astringent wine that mellows with ageing in wooden casks and in the bottle. In the Médoc all wine made from Cabernet Sauvignon is blended with Cabernet franc and Merlot, which mature earlier, before bottling. Many of the great wines of the Cape are made from 100 per cent Cabernet Sauvignon grapes. To have the wine from these grapes fit for the table after only a few years' maturation requires skilled cellar techniques. Blends with Merlot and Cabernet franc are becoming more popular.

CARIGNAN

The most widely cultivated red variety in the world, Carignan is well adapted to harsh, dry and hot conditions and grows vigorously, even in very sandy soils. In North Africa, Carignan has demonstrated an ability to grow and produce well where no other *Vitis vinifera* variety can grow. This ability makes Carignan suited to the conditions on some Cape Estates. Carignan bears heavily, and should be grown as bush vines and pruned short. The grapes tend to be high in acidity, especially in cooler areas where Carignan produces harsh, astringent wine. Under warm conditions, the acidity falls to reasonable levels, producing wine with a ruby red colour and a natural equilibrium.

CINSAUT

Grown on dry hills in France, without irrigation, Cinsaut is capable of producing wines of reasonable quality, when blended with Carignan and Grénache. However, it has the capacity to produce large volumes of low quality wine and this has destroyed any reputation the variety may have had in the Cape. Most Estates producing red wines had extensive Cinsaut vineyards in the early '70s, but most of these have since been uprooted. Even under optimum conditions in South Africa, Cinsaut loses acidity as it approaches ripeness and provides the cellar master with difficulties in trying to make an acceptable wine.

GAMAY

The second cultivar of Burgundy, Gamay has not proved popular in South Africa, being grown on only one Estate, and is seen as a poor alternative to Pinot noir. In France, Gamay's best wine is Beaujolais. It ripens early in the season and makes a dark-coloured wine that matures quickly.

MERLOT

Most famous for its role in the production of the best clarets, Merlot is the second most important variety after Cabernet Sauvignon in the blends of some of these wines, and is the dominant partner in others. It is becoming increasingly popular in California, Australia and South Africa because it bears more grapes than Cabernet Sauvignon, produces wine with a distinctive flavour and a fine Cabernet aroma, and blends harmoniously with Cabernet Sauvignon to produce early-drinkable wines that also have a capacity to last.

PINOTAGE

A hybrid descendant of Pinot noir and Cinsaut, the grapes of this variety ripen early under South African conditions, reaching high sugar levels with high acidity. Resistant to most vine diseases (except oidium), with a capacity to produce large volumes of wine with respectable depth of colour, Pinotage became particularly popular with South African growers during the '70s. But a degree of consumer resistance to the lesser Pinotage wines has seen a changed attitude toward this cultivar. At best, the wines are excellent, but its very distinctive flavour often tends to dominate blends. Pinotage has been shown to be useful in the making of port.

PINOT NOIR

Pinot noir is the noble variety used in the production of the great Burgundian reds, and plays an important part in the production of white wine for Champagne. The grapes ripen with a high proportion of sugar, and the wine has a beautiful ruby red colour. However, South African cellar masters have found difficulty in making richly coloured red wines with a significant depth of delicate flavour. Major problems apear to be caused by over-production, irrigation and the youth of the vineyards. Growers are also still searching for the right micro-climate. But very promising wines have been made, and expectations are still high.

SHIRAZ

In France, this cultivar is known as Syrah and is used to make the famous Rhône wines of Hermitage (not to be confused with the South African term Hermitage, which was formerly used to describe the Cinsaut vine). Shiraz bears well when pruned long, ripens in late mid-season and produces a rich, full-bodied wine, much in demand in South Africa. It is, like Cabernet Sauvignon, a difficult grape from which to make wine, requiring great care in cellar technique, but when properly handled, Shiraz is capable of making some of the best wines in the country.

SOUZÃO

Originally a port variety, this cultivar has proved to make medium to high-quality red table wine. It is very susceptible to wind damage and sunburn and so is not widely planted in the Cape wine country. It bears well and ripens mid-season; grapes have a high acid and sugar content and rich dark colour, which together make a colourful, well-flavoured wine. Souzão is popular in the Malmesbury area, where it can be trellised without great fear of wind damage. Small vineyards have been planted on some Estates to provide deep coloured wines to blend with lighter styles.

TINTA BAROCCA

Imported for the production of ports, this variety produces copious quantities of medium quality table wine and some port. It grows well and is resistant to most vineyard hazards, but lacks delicacy of character. It is most popular in the Swartland area, where it produces generously, reaching high sugar levels, while retaining sufficient acidity.

ZINFANDEL

This high quality variety is not a strong-growing vine and should not be planted on poor soils, especially sandy soils, where it easily becomes sunburnt. Best results are achieved from bush vines grown without irrigation, producing grapes with high acidity. The wine is best when made soft and light to make the most of the fruity flavour.

Cultivars/Varieties

The Climatic Factor

It is generally accepted that wine-making technique and the genetic qualities of the vine (the cultivar's inherent quality) are the two most powerful influences on the quality of wine. However, when one uses first-class cellar technology and works with the accepted classic varieties, only climate and soil are likely to provide significant differences in quality. Soil has long been viewed by Europeans as the cause of specific qualities in their wines, but this has been generally discounted in recent years. The chief role of soil seems to be the provision of a balanced growing medium for the vine to produce sufficient foliage and crop to enable it to ripen the crop to the required degree of maturity.

In 1935 specialists in California began their studies to discover the factors with the greatest influence on the quality of wine. In 1944 Professors Amerine and Winkler concluded that the temperature of the surroundings in which the vine grew and ripened its grapes was the only climatic factor to have a marked effect on the composition of the grape fruit at peak ripeness. They developed the heat summation method of classifying vine-growing districts into regions of comparative average temperature over the vine's seven-month growing period.

They set 50 °F (10 °C) as their base figure, as it is generally agreed that there is no shoot growth below this mean (whole day average) temperature. During the seven-month annual season of the vine (April to October in the northern hemisphere and September to March in the southern hemisphere) one is able to summarise the heat available to the vine by adding the mean temperature for each day above 10 °C to produce a total, expressed as degree-days. For example, if a weather station at Groot Constantia has a mean temperature of 20 °C for the whole of January, it will have 310 degree-days in that month.

Using this technique Amerine and Winkler were able to discover that there were regions in California with heat summation totals similar to those of Bordeaux, Burgundy, Champagne and the German top-quality areas, as well as others that correspond to warmer regions in Spain, Portugal, Italy and Algeria.

In 1974 Ernst le Roux, a South African viticulturist, attempted to lay a similar scale over the Cape wine-growing districts, but found insufficient detailed temperature information to delineate areas with reliability. Since then, a number of new weather stations have been installed in potential and existing wine-growing districts. However, the evidence presented to date can only be seen as an indication. Thirty consecutive years is the accepted statistical minimum for the drawing of definite conclusions.

Temperature is able to play such an important role because certain acids and flavour compounds in grapes are volatile and are lost from the juice during periods of greater heat.

Because of South Africa's geographical position with Cape Agulhas at less than 35 ° of southerly latitude (quality wines are made in Germany north of 50° latitude) the average temperatures of the present growing regions are somewhat warmer than Europe, the west coast of the United States and southerly regions in Australia, New Zealand and Chile.

Because grape-growing areas in South Africa have traditionally been chosen for their capacity to provide a large volume of fruit of moderate quality, the colder regions, with resultant lower production, have been largely ignored. This has meant that almost all South Africa's grapes are grown in regions that qualify as Region III, according to the Winkler classification provided below. This practice has been entrenched by the historical fixing of grape-growing quotas to vineyards. Nevertheless, South Africa has several districts that qualify as Region I, and though no grapes may be grown for wine production in those areas at the time of writing, it is hoped that this situation will change in the future and allow us to discover whether there are indeed better areas or not.

Further experience and research will show us just how the shorter day length in South African vineyards (when heat summation figures are similar) and other uniquely local factors have an influence on the quality of the wines of this country. Cabernet Sauvignon has already demonstrated that it is unable to ripen its grapes properly in many Region II and III vineyards in this country. Thus care must be taken not to read too much into temperature alone.

The Winkler classification
Region I   less than 1389 degree-days
Region II   1389 – 1667 degree-days
Region III   1668 – 1944 degree-days
Region IV   1945 – 2222 degree-days
Region V   more than 2222 degree-days

SOME SPECIFIC WINKLER MEASUREMENTS

These figures have been developed from temperature measurements taken at weather stations in some existing and potential wine grape growing districts. They are specific and apply only to the conditions at that weather station and may not apply to vineyards growing nearby. However, they do give an idea of the relative warmth of some of the Cape regions. For comparison we have included a few random figures from European and American weather stations.

WEATHER STATION	° DAYS	AVE. °C PER DAY IN RIPENING SEASON JAN	FEB	MAR
Reims, France	1011			
Bordeaux, France	1328			
Grabouw (Overberg)	1486	19,3	19,55	18,55
Napa, USA	1600			
Hermanus (Walker Bay)	1609	19,6	19,95	18,9
Constantia (Constantia)	1720	20	20,6	19,75
Firgrove (Stellenbosch)	1756	20,6	21,35	20
Koelenhof (Stellenbosch)	1898	21,45	22,15	20,7
Durbanville (Durbanville)	1906	21,4	22,4	20,85
Groot Drakenstein (Paarl)	1967	21,5	22,1	21
Ashton (Robertson)	1995	22	22,4	20,75
Franschhoek (Paarl)	2014	21,75	22,4	21,1
Rawsonville (Worcester)	2032	22,35	23	21,25
Wolseley (Worcester)	2104	22,8	23,0	21,6
Oudtshoorn (Klein Karoo)	2143	23,05	23,4	21,55
Robertson (Robertson)	2164	23,1	22,9	21,5
Tulbagh (Tulbagh)	2230	23,45	24,05	22,2
Calitzdorp (Klein Karoo)	2232	22,3	23,4	21,85
Dal Josaphat (Paarl)	2280	23,5	24,5	22,6

The South African Wine Industry

CONSTANTIA

Constantia

Groot Constantia

Constantia was subdivided into many small areas. Several of these have been consolidated into Groot Constantia.

Constantia is one of the most famous wine farms in the world. Wine bearing the Constantia label was served to kings and queens throughout Europe during the eighteenth and nineteenth centuries. In Jane Austen's novel *Sense and Sensibility*, written in 1811, the wine is referred to as "the finest old Constantia wine . . . that ever was tasted", and recommended for its "healing powers on a disappointed heart". And it was indeed Constantia wine that comforted Napoleon's last days, together with champagne and claret from France. Such was Napoleon's preference for Constantia that the English Governor of St. Helena, where Napoleon was imprisoned, complained of the expense involved in keeping the ex-Emperor supplied with his South African wine.

Though the property has not always been a wine farm, Constantia became an important part of South Africa's wine industry as early as 1680, through the energy of Simon van der Stel, and is still an important wine Estate today, 300 years later. Born in Mauritius, the son of the Governor of that colony and of an Indian woman, Monica of the Coast, Simon van der Stel came to the Cape Colony as Commander (later Governor) in 1679. He left his wife, four sons and daughter in Holland, and was never to see his wife again. Each of his sons later lived for some time in the Cape. Within weeks of arrival in the country, he founded Stellenbosch. He opened up areas in the Paarl and Stellenbosch basins that were later to be farmed by the Huguenots.

At a time when the Dutch East India Company considered that the Cape was simply a food and provision post for ships on their way to and from the Dutch East Indies, Simon van der Stel opened up large tracts of land for farming and settlement. He was the first Governor of the Cape to take an active interest in the development of what was eventually to become South Africa. He personally explored as far north as Namaqualand, to where the present O'Okiep copper mine is sited. He encouraged agriculture in general and the planting of oaks in particular. South Africa must be indebted to him for the many thousands of huge old oak trees that are to be found today throughout the Cape's Western Province.

Simon van der Stel was a good Governor and administrator, but he became acquisitive and as time passed he began to do less for farmers and colonists and more for himself. He had chosen Constantia for himself before he was appointed Governor, and such was his influence that, though

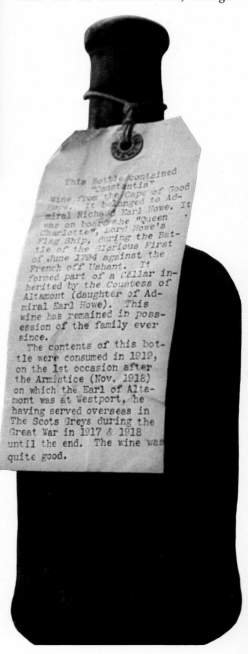

the average private grant of farming land was some 50 hectares, Constantia covered 750 hectares. Later, while Governor, he granted himself the grazing rights for virtually the rest of the Cape peninsula, with land amounting to some 8 500 hectares. On his death in 1712 these grazing rights expired, and eventually the huge farm of Constantia was split up into eight separate properties, of which Groot Constantia was one.

It has been generally believed that Simon van der Stel knew little about agriculture and still less about vineyards. Yet while Commander of the Colony he had soil tests made in the vicinity of the settlement and chose Constantia, on the eastern slopes of the Table Mountain range, facing False Bay, for his own property. Whether by skill or by accident, he chose some of the finest vine-growing soil in South Africa. He was granted Constantia in 1685 and built his Manor House on the property in 1692. He planted the first vines on the Estate before 1690, and made wine for sale in the Colony and also for export to Europe. The types of vines planted by Van der Stel, the quality of his wine and his influence on vine growing in the Western Province are unknown. Almost a century was to pass before Constantia was to become a household word in the palaces and mansions of Europe. After Van der Stel's death and the fragmentation of the original Constantia, the part that contained the original homestead and is now known as Groot Constantia was reduced to an area of some one hundred hectares. Wine from this farm was shipped by its various owners throughout the eighteenth century, but seldom in quantities of more than fifteen tonnes a year.

The soil is rich and loamy, derived from Table Mountain sandstone, and its composition bears no relation to the granitic and shale soils of Stellenbosch and Paarl. In fact some of the Estate's greatest red wines have been made on the lower slopes of the farm,

Wine was made, stored and sold in this original cellar. Today it contains a museum and is used for wine sales.

This relief sculpture in the pediment of Groot Constantia's wine cellar is the work of Anton Anreith, South Africa's first great artist. It was executed in 1791.

The architect Thibault was responsible for the style of the manor house, now a national monument and museum.

Constantia

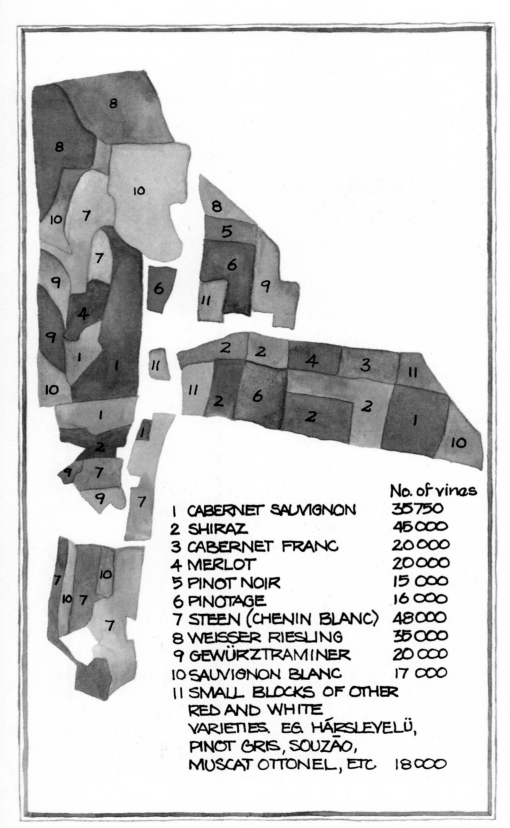

where the sandy composition of the soil is even more pronounced. Irrigation is not used in the vineyards at any time; the vines and grapes gain their water from the rains that come across the mountain from the north-west. Because of Groot Constantia's position on the eastern slopes of the 1 000-metre-high Table Mountain range, it has little afternoon sun, and the relatively low temperatures slow down the rate of ripening. The harvesting period begins at the end of February (by which time some Estates farther north have been picking grapes for four weeks) and ends in late April.

Hendrik Cloete, descendant of one of the first free burghers, or citizens, of the Cape Colony, bought Groot Constantia in 1778. A superb description of Hendrik Cloete and his prowess as a wine maker is found in C. de Bosdari's *Wines of the Cape*:

"Hendrik Cloete must have been a remarkable man and a very good wine farmer. The place (Groot Constantia) had been neglected; but he soon put matters right in the vineyard and outside it, repairing the old wine cellars and building a big new one, and enlarging and remodelling the original Van der Stel house. He made Constantia into one of the world's great wines, and even corresponded with Crowned Heads about it. The red wine was in all probability a blend of Pontac, Red Muscadel and Red Muscat de Frontignan; there was also a white, which was rated as highly as the red. The precise details of the blends are lost. But it is known that harvesting was postponed until the grapes were over-ripe, so that the quantity was much reduced; it was therefore probably a sweet, rich, full wine of delicious flavour and pronounced muscat fragrance."

Groot Constantia was bought by the British Colonial Government from Jakob Pieter Cloete on 19 October 1885, and has remained in Government hands ever since. As the quality of and demand for Constantia wine had steadily decreased over the years, the Colonial Government first established an agricultural experimental farm on the Estate. On this farm, young men from all over the Cape Province learnt better methods of farming with sheep, fruit and vines. In 1926 the South African Government established a private company named Groot Constantia Estate (Pty.) Limited and ran the farm as an Estate wine farm, planting the varieties popular at that time. These vines produced sweet white table wines, and fortified sherry and port wines.

In 1943 the Government leased the farm to the Bertrams wine organisation, which has long been connected with the South African wine industry. On Constantia Bertrams made sweet white wine for sale in Cape Town and dry red wine for export purposes. These wines were made from the old port-type and sherry-type vines, plus Steen and Cinsaut. In 1957 Bertrams's lease expired and it was not renewed by the South African Government. Once again, the Groot Constantia Estate began to make its own wines. The farm was run by the Department of

(above) Groot Constantia has a very moderate climate, without a wide temperature range or other extremes.

The vineyards are not irrigated and the high humidity level does not vary throughout the summer.

Groot Constantia

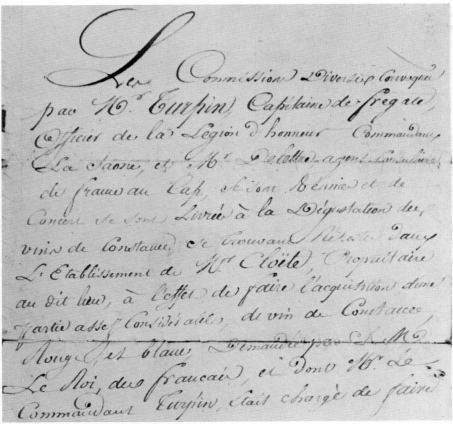

Authority given to a Mr. Turpin to select Groot Constantia wines for the King of France.

Over six decades, demand for Groot Constantia wines has kept pace with the rising prices and limited supply.

Agriculture until the formation of the Groot Constantia Control Board, an independent authority created by the Groot Constantia Estate Act of 1975.

The Estate is now operated as an independent entity by this Board, which is answerable to the Minister of Agriculture.

The Control Board brought together a management team, consisting of Johan Neethling, André Meiring and Neil Ellis, to direct the planned improvements to the farm. Johan Neethling died soon after the team's first vintage and was replaced by Koos Stofberg. Neil Ellis left the Estate after the 1982 harvest to take control of the Zevenwacht Estate at Kuils River and was replaced by Pieter du Toit.

In the re-establishment of the glory of Groot Constantia the Board set two main objectives. First as much of the original Constantia as possible was to be brought back into the Groot Constantia Estate, and secondly the vineyards and cellar on the farm were to be replanned and equipped to bring Groot Constantia back to the forefront of the South African wine industry. In 1975 and 1976, three pieces of ground, once part of Constantia, were purchased and incorporated into the Estate. The house 'Hoop op Constantia' and the land surrounding, as well as part of the historic Nova Constantia Estate were threatened by housing development schemes and were bought by the Control Board for vineyard extension. An adjoining piece of forestry land, high on the slopes of the Constantiaberg, was bought to bring the area of land to 150 hectares.

At the time of the Control Board's move into Estate management, Groot Constantia was known to be a producer of limited quantities of Cabernet Sauvignon, Shiraz and Pinotage wines. About five per cent of the Estate's vineyards were planted with Steen (Chenin blanc), and these were used to produce a small quantity of semi-sweet white wine.

As none of the additional land was planted with vineyards, it was decided to increase the Estate's white wine

As on most Cape wine Estates, Constantia has a permanent force of labourers, supplemented during harvesting.

Maintained as an eighteenth-century museum, the manor house is even used by educational authorities to teach history.

production by planting quality white varieties on these differing soils, higher and lower than the established part of the farm where red varieties had proved their suitability over the previous twenty years. A further aspect of the replanting programme involved the removal of the old red vines from the central part of the farm in a phased three-part plan and their replacement with superior, virus-free stock of similar varieties.

Groot Constantia is considerably further south than most of South Africa's premier wine growing districts and has a longer and later ripening season than farms in warmer areas. Traditionally, Cabernet Sauvignon, a late variety, reaches peak ripeness in the weeks before winter at Groot Constantia. Under the influence of a cooler summer or the onset of an earlier winter, the point of peak ripeness was not reached before the acid in the grapes began to drop and harvesting was therefore begun. The new vineyards, with their better-prepared soils and improved planting material, reach peak ripeness more easily, reducing the likelihood of 'unripe' wines and allowing the optimum balance between sugar and acid to be reached more often. The Estate's reputation for fine, clean, medium-bodied red wines will probably be reinforced once the produce of the replanted vineyards comes into the cellar.

With the Estate's increasing emphasis on blends, the cultivar wines of Cabernet Sauvignon, Shiraz, Pinotage and Pinot Noir have been joined by two important wines. It is likely that Heerenrood, a Shiraz-based blend, and the new Bordeaux blend of Cabernet Sauvignon, Cabernet Franc and Merlot, will become Groot Constantia's premium red wines. Pinotage assists in the production of an interesting, steely-tasting rosé wine, and Shiraz, which may be the most valuable red variety on the farm, is also used as the base of an Estate port.

The new white varieties are further advanced than the new red, having been planted earlier. First among these appear to be Weisser Riesling and Sauvignon blanc, making full, fragrant wines with great depth of flavour. Steen, as always on this farm, produces a wine of clean, sharp intensity that is well suited to the few grammes of sugar left over from the harvest, with which it is bottled.

Groot Constantia is surrounded by suburban development, but is protected from subdivision by legislation.

Groot Constantia produces slightly more white wine than red, and this will remain the pattern for the foreseeable future. The bulk of the white wine production will be used to slake the thirst of the Estate's summer visitors, while the reds, together with the better white wines, will be matured for the connoisseur market.

In 1982, 30 000 cases of wine were bottled at Groot Constantia (a great advance in production over 1975, when the figure was 10 000), with a significant effort being made in the maintenance of quality.

Wine is for sale on the Estate.

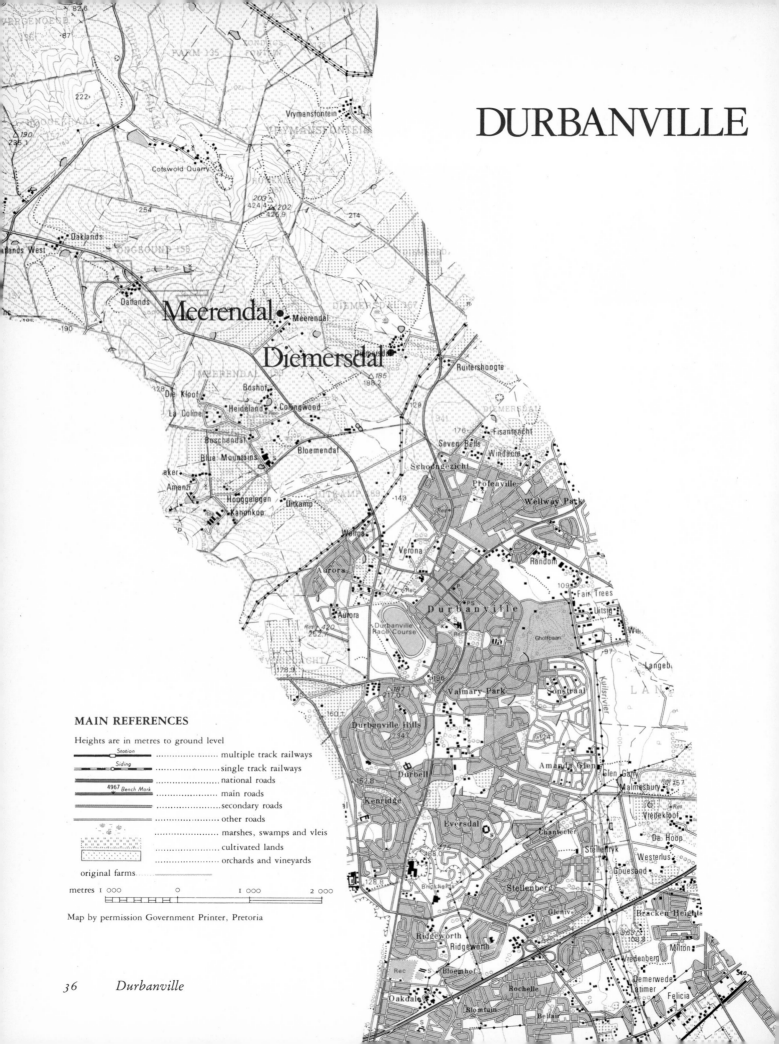

36 *Durbanville*

Diemersdal

Before the recent appreciation of quality wines, Durbanville was not an important wine area.

Looking north-east across the valley at the head of Table Bay, from the site of the first European colony, one can see an eroded range of hills, known as the Tygerberg, where the Diemersdal and Meerendal wine Estates can be found today. Once the haunt of leopards preying on stock grazing on the grassland, today the northern suburbs of Cape Town are spreading across the slopes, encroaching on the rich, red soil that produces some of South Africa's best red wines.

The valley beyond the hills was used to fatten cattle and sheep from the time of the first settlement, but, lacking a river or a reliable source of summer water, was left out of consideration in the search for the Cape's early agricultural settlements.

In 1698, Hendrik Sneewind was granted 66 hectares of land on the north-facing slopes of the most northern Tygerberg Hills. Sneewind's son-in-law, Abraham Diemer, must have run the farm when father-in-law became too old to do so, because the farm became known as Diemersdal, even though Diemer, who owned two other properties closer to Stellenbosch, was never the official owner. In 1714 the farm was transferred from Sneewind's name to that of J. van der Westhuyse.

Like the adjoining farm Meerendal, Diemersdal has steeply sloping land with stony soil, lower slopes with red soil based on a layer of clay, and sandy loam soils in the valley at the foot of the hills. Diemersdal has been a mixed farm since Hendrik Sneewind's time, grazing cattle, growing vines and fields of grain. Most of the vineyards are planted on the strip of rich, red soil that extends at a fairly constant level from the suburbs of Durbanville through Meerendal and ends within the boundaries of Diemersdal. The other soils are used to graze dairy cattle and sheep and to grow wheat. Where the red Durbanville soil ends, a ridge of Hutton-type soil begins, and this is also used for the planting of vineyards. The rainfall in this area generally ranges from 450 to 520 mm per year, which is sufficient to grow sturdy vines and ripen a crop of grapes. Because of the absence of a regular source of summer water, none of the vineyards is irrigated.

Diemersdal was first settled in 1692. The main gable bears this date as well as the date of restoration.

The semi-circle strip of red vineyard soil near Durbanville ends on the Diemersdal Estate.

The low density planting philosophy, with widely spaced rows, can be seen in this Pinot noir vineyard.

Diemersdal's vineyards are planted on gently sloping ground predominantly facing north, and the red soil appears to be slightly deeper on this Estate than it is on Meerendal. The layer of clay under the Durbanville red soil allows grapes to reach high levels of sugar without causing the vine discomfort and a subsequent loss of acidity in the grapes. The wind blows strongly from the south-east in summer and lowers temperatures during the late afternoon and evening. Diemersdal is only about twenty kilometres from the west coast and the cold Atlantic Ocean, and receives

Diemersdal

supplementary summer breezes from this direction that bring moisture and cause considerable overnight precipitation on the vineyards in summer.

Tienie Louw bought Diemersdal soon after the beginning of this century and maintained the dairy, wheat fields and vineyards he found there. He increased the vineyards to 22 hectares during his lifetime and these vines provided the farm with its major source of income. Tienie was more of a businessman than a farmer, and had a vinegar factory in Salt River, Cape Town, that obtained some of its raw material from his own vineyards. The vines were also used to make dry red and white wines, some of which were used for distillation. Tienie was an enthusiastic builder and established a cellar, a dairy and storage buildings for his grain, animal feed and equipment around the 1698 manor house, giving Diemersdal the appearance of a village.

When Tienie Louw died in 1942 Diemersdal was inherited by his seventeen-year-old son, Matthys. Though it remained a mixed farm, and the sheep, cattle and wheat continued to be important elements of production, Matthys began to capitalise on the bounty that nature had provided in the soil. He planted new vines during most of his years on the farm, increasing the area under vines from 22 to 170 hectares. He planted early, mid- and late-season varieties in the red soil and early-ripening varieties in the Hutton soils. Soon after taking charge at Diemersdal, he started supplying Dr. Charles Niehaus of the KWV with red wine and sherries made in the farm cellar. When KWV sherry sales decreased, Matthys concentrated on the production of red wines, with the produce of his white variety vineyards being supplied to the KWV cellar in Paarl at grape stage. Matthys was one of the first Estate farmers to produce Pinot Noir and Cabernet Franc wines in the mid '70s.

When neighbouring Meerendal stopped supplying the KWV with white grapes and red wine in 1974, the co-operative asked Matthys to plant more vineyards of both red and white varieties to enable them to maintain their level of valued Durbanville origin wines. Matthys planted more Shiraz and more Steen, the well-ripened produce of which was to be used as an important element in KWV's Late Harvest wine, a favourite among co-operative members.

During Matthys's last few years on the farm, he extended the white variety vineyards with Kerner and Weisser Riesling and added Merlot to the list of red vineyards. In 1982 Matthys died and was succeeded by his two sons, Tienie and Beyers.

At Diemersdal a new vineyard is prepared by using a large tractor to sub-soil, or deep-plough, reaching sufficient depth to disturb the surface of the clay. The red soil has a balanced pH analysis and is given an infusion of organic fertiliser to improve the texture and nitrogen content. The Hutton soils require additional lime to lift the pH to required levels. The new vineyards were planted in the

The Diemersdal vineyards are grown on gently sloping ground and are surrounded by wheat fields.

Though Diemersdal harvest red and white varieties, only the red grapes are crushed and fermented on the Estate.

1. SAUVIGNON BLANC
2. STEEN
3. CABERNET SAUVIGNON
4. PINOTAGE
5. CABERNET FRANC
6. KERNER
7. WEISSER RIESLING
8. TINTA BAROCCA
9. PINOT NOIR
10. MERLOT
11. SHIRAZ

same year that the old vineyards were removed. The vines are planted 1,2 metres apart with rows 3,1 metres apart, providing a vine population of 2 250 per hectare. Though current trends in South Africa favour more dense vineyards, Tienie uses one set of tractors for both wheat fields and vineyards and therefore requires wide passage between the rows for cultivation. The space between the rows is kept clean of growth by tilling the soil each season to a depth of one metre, and vegetation between the vines in each row is removed by hand. Organic fertiliser and phosphate are mixed with the newly-turned soil.

The long, slow ripening period and the acid content of the grapes influence the cellar toward the production of full-bodied wines, and Tienie tries to pick his grapes at full-ripe stage. The red grapes are crushed in the Estate cellar, while the white grapes are sent to Paarl. The red must, with skins and must mixed, is cooled to 24 °C and fermentation is allowed to begin spontaneously, using the vineyards' native yeast, in open concrete tanks. More than half the sugar present at crushing is allowed to ferment out before the skins are removed for pressing. About two-thirds of the pressed must is returned to the liquid fermenting in closed concrete tanks, where all the sugar is removed. The wines are racked to remove most of the lees and allowed to await malolactic fermentation. After the secondary fermentation the wines are cleaned and transported to the KWV for maturation in oak. The matured wines are exported and used to fill out the KWV wines produced for sale to members. Small quantities of each of the Diemersdal reds have been retained by the KWV for possible future sale under the Estate label.

Tienie Louw plans to install a small wood-maturation cellar at Diemersdal and to bottle a limited quantity of the Estate produce for sale to the general public.

At present, wine is not for sale on the Estate.

Diemersdal is still today a mixed farm with wheat fields, pastures and a commercial dairy.

Diemersdal

Meerendal

One of the many Dutch East India Company employees to be declared a free citizen and granted farming land, Jan Meerland, who worked with the Company's stores of timber and building materials, was in 1702 granted 50 hectares of land, alongside Hendrik Sneewind's Diemersdal property, by Willem Adriaan van der Stel. Meerland's property, Meerendal, in the south and east-facing folds of the Tygerberg Hills near Durbanville, is today a mixed farm with vineyards, sheep and wheat fields, and is best known for the quality of the red wines produced in the Estate cellar.

Jan Meerland, while working in the Company store, was assigned to a work party which was given the task of salvaging the wrecks of the *Oosterland* and the *Waddinxveen*, ships that had been driven on to the Table Bay shore during the winter of 1697. The *Oosterland* is known to have brought a party of Huguenot refugees to the Cape in April 1688, and was presumably wrecked after several subsequent voyages to the Cape.

Meerland did not spend much time on his Tygerberg farm. He was closely associated with both Henning Hüsing of Meerlust and Adam Tas of

Looking across the Meerendal vineyards on the famous red soil, towards the wheat fields on sandy soils.

Libertas, and was involved in the agitation against their common benefactor and Governor, W. A. van der Stel, and in the events leading to the revolution of 1706. Van der Stel had Meerland, Hüsing and three others declared to be ringleaders of a revolt and banned them from the colony. They were sent back to Holland under escort, but unknown to Van der Stel members of the military escorting the pioneers were sympathetic to their cause, and allowed them to carry a letter written by Tas to the Council of Seventeen denouncing the conduct of the administration of the colony. As a result, Van der Stel and his associates were removed from their posts. Unfortunately, Meerland died during the voyage to Holland.

Though agriculture was made difficult by the lack of summer water, the rich soils and the proximity to Cape Town encouraged the development of mixed farming on Meerendal.

When William Starke bought Meer-

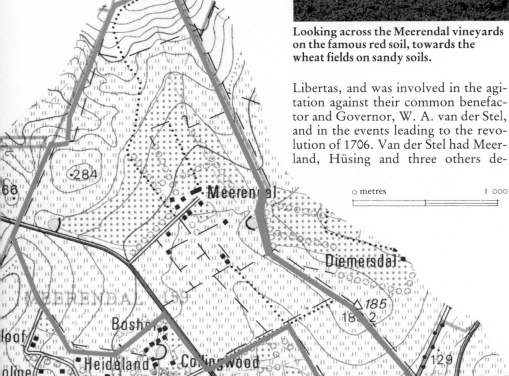

Stock farmer William Starke bought the grazing property Meerendal in 1930, planted vineyards and started making quality wines.

Meerendal produces excellent white grapes. These young Sauvignon blanc vines are the first of many.

The distinctive colour and flavour of Meerendal Pinotage is obtained from the skins, here being punched through the fermenting must.

endal in 1929, 28 hectares were planted with vines and the rest of the 400 hectare property was used to grow wheat and graze cattle and sheep. William Starke had been raised in the Muldersvlei area between Stellenbosch and Paarl, and before buying Meerendal was the farm manager of an agricultural college. He married a daughter of the famous Faure wine-producing family, and extended the vineyards on his new farm with the first block of Shiraz vines in 1932.

The Faure family were able to introduce William to wine buyers at the KWV, and Meerendal Shiraz, fermented in the farm cellar, was provided to the KWV for blending into the red wines destined for the export market. The first indication that something unusual was coming from the vineyards on Meerendal's red soil was when Dr. Charles Niehaus, on his appointment to the post of wine expert at the KWV after the Second World War, revised the standard blend of the KWV's premium red wine, Roodeberg, and made Meerendal Shiraz the base of the new blend. Meerendal also had a port-variety vineyard with mixed Cornifesto, Tinta Barocca and Souzão vines, and the Shiraz and port supplied to the KWV were the farm's two major products during William's time.

Koosie, William's son, joined his father in 1948, and took charge in 1952. This coincided with the first commercial plantings of Pinotage, a hybrid developed by Dr. Perold of the KWV, and Koosie was encouraged to plant this untried red variety by Dr. Niehaus. From 1956 until 1974 Koosie continued to supply the KWV with two red wines and port made in the Meerendal cellar, as well as the crop of grapes from Meerendal's Steen vineyards. The white variety was planted by William and retained by Koosie to fill a harvesting gap between Pinotage and Shiraz.

Throughout their association, the KWV had used Meerendal's wines to produce their premium products, which were sold to members of the Co-operative and exported to overseas markets. When interest began to develop in Estate wines in the early '70s, the KWV bottled small quantities of wine under the Meerendal label and made these available at first only to their members. Because the KWV is prevented, by charter, from marketing wine to the South African public, Meerendal's wines were rarely tasted by local wine lovers. In 1974, Koosie Starke switched allegiance to a Stellenbosch wine wholesaling company and the first Meerendal Shiraz and Pinotage wines available to the general public made their appearance during 1975.

The Meerendal vineyards and wheat fields are on the slopes and in the valley at the foot of one of the northernmost hills in the Tygerberg. The vineyards are situated on south and east-facing slopes in the lower section of the hills, and like the other

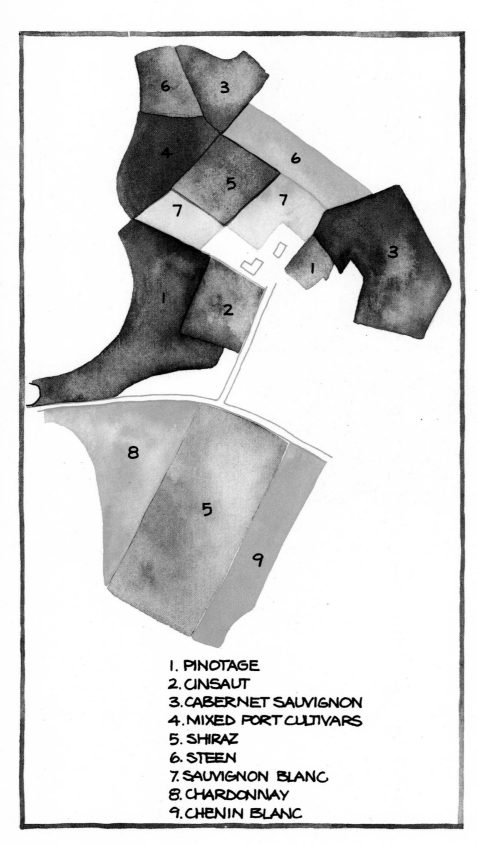

1. PINOTAGE
2. CINSAUT
3. CABERNET SAUVIGNON
4. MIXED PORT CULTIVARS
5. SHIRAZ
6. STEEN
7. SAUVIGNON BLANC
8. CHARDONNAY
9. CHENIN BLANC

Germanic-influenced wood carving can be seen throughout the Cape Wine industry. This is the work of Alec Serra.

Koosie Starke, Meerendal's vintner, is a specialist red wine producer.

Meerendal farming activities their location is determined by the soil. Meerendal has three types of soil: from a point roughly half-way up the hill to the crest the soil is stony and difficult to cultivate; below the stony ground a layer of rich, red soil based on clay extends to the floor of the valley, where in turn sandy loam soils predominate. Koosie uses the upper portion of the farm for grazing, the lower section on the level sandy soils for wheat and the red soils for vineyards. This heavy soil, with its layer of clay, holds sufficient moisture for the vineyards to grow and ripen grapes in even the driest summer, and may be seen, during the period of preparation for wheat fields, running around the basin of hills, at roughly a constant elevation from the northwest suburbs of Durbanville, through

Meerendal's red soils have sufficient fertility for growth and enough moisture to ripen grapes fully.

Meerendal and into adjoining Diemersdal, where it ends.

The benefits provided by this unique soil structure are enhanced by the pattern of summer winds that blow through these hills, sometimes causing damage to the vines, but always bringing cool air to the vines straining to ripen their crop and helping them to replace volatile elements lost in the heat of the day. Summer nights are cool in Durbanville and there is heavy precipitation of dew, providing moderating factors that tend to make up for the rather intense heat at noon on most summer days.

Vines are capable of attaining high levels of sugar in their grapes without dramatic decreases in acidity. Meerendal receives an average of 550 mm of rain each year, and none of the vineyards is irrigated. The grapes produced in the vineyards are best suited to the production of full-bodied wines. Koosie Starke changed his cellar technique to try to modify the style of wine to suit the preferences of the market during the '70s, but was subsequently forced further to modify the new technique and once again produce a full-bodied wine. The grapes from the red varieties are picked at full-ripe stage. After crushing, the mash containing juice and skins is cooled and inoculated with dry yeast to start fermentation. Until 1977 Koosie used the farm's native yeast to start spontaneous fermentation, but changed to dry yeast to start immediate activity and to get the natural protection of the CO_2 given off the must. He also began to separate the skins and fermenting must at an earlier stage than had been Meerendal's traditional practice. He has reverted to fermenting with the skins down to 6° Balling before separation to get greater flavour extraction, and believes that the cooler temperatures (slower fermentation rates) being used in recent years are producing wine with a more delicate and appealing flavour.

In traditional style, Koosie ferments the red musts in open concrete tanks until the skins are removed, and completes the fermentation in closed tanks. The wine is then pumped off the sediment of lees in the base of the tank and given a small quantity of SO_2 for protection. Spontaneous malolactic fermentation normally begins within a week. All the Estate's red wines are then transported to Stellenbosch for maturation in oak. Though only Shiraz and Pinotage wines are sold under the Meerendal label, the Estate also has a Cabernet Sauvignon vineyard. These vines, planted in the early '70s, to see what this classic cultivar would be able to produce on Meerendal's renowned soils, were unfortunately grafted on to inferior rootstock and the vineyard became infested with phylloxera. Koosie has implemented a gradual replacement programme, replanting the vineyard in stages with healthy vines. The future of this cultivar on Meerendal looks promising as Koosie has made some excellent Cabernet Sauvignon wines from the stricken vines.

However, Koosie has been encouraged by his Stellenbosch merchant partner to increase his production of white grapes. Like his father, he has continued to make only red wine in the Estate cellar and the white grapes are supplied, uncrushed, to the cellar in Stellenbosch. He has maintained his vineyards of Steen, and has planted blocks of Sauvignon blanc and Chardonnay, raising the number of hectares under vine to 125, with white varieties occupying about 40 per cent of the space.

The Meerendal red wines, matured in Stellenbosch, are normally bottled when two years old and released after a further year of bottle maturation. They are characteristically deep in colour, rich in flavour, and surprisingly soft in tannins, and though able to undergo long maturation with beneficial results, are pleasantly round to the taste while young.

Wine is not for sale on the Estate.

LITTLE KAROO

MAIN REFERENCES

Heights are in English feet to ground level

- multiple track railways
- single track railways
- national roads
- main roads
- secondary roads
- other roads
- marshes, swamps and vleis
- cultivated lands
- orchards and vineyards
- original farms

Map by permission Government Printer, Pretoria

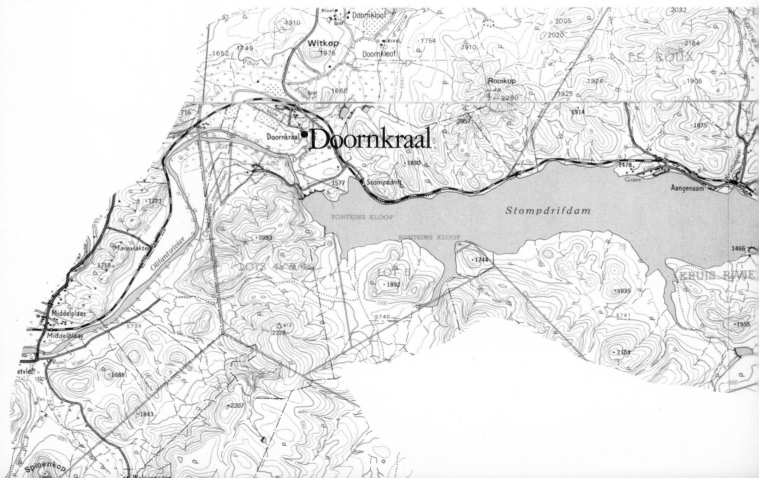

Boplaas

Two small Calitzdorp vineyards, four kilometres apart, supply the Boplaas cellar with grapes to make dry and sweet wines. The cellar stands on 15 hectares of vineyard and orchard land beside the Nel's River, on the edge of town. The majority of the Estate grapes are grown on a 35 hectare property, called Welgeluk, just outside the municipal boundary.

Calitzdorp was one of the earliest inland settlements in the Cape. Though it took more than six days to travel to the Little Karoo from Cape Town by ox-wagon in those pioneering days, the valley offered fertile soil, two rivers that contained water at all times of the year and extensive grazing lands to the east and north. To provide stock farmers with some security the loan farm system had been instituted by the Dutch East India Company in 1703, and all the land around Calitzdorp and further east into the Little Karoo was settled by families renting their properties from the administration for initially twelve and later twenty-four stuivers per year.

All these loan farms were originally about 5 000 hectares. Later they were reduced to 2 500 hectares and were roughly measured out by using the distance that a horse could walk in half an hour from a central point (usually where the settler had already built a house) as the means of determining the boundary. Subdivision of such a farm was not permitted.

The first loan farm recorded in the Little Karoo was settled by J. J. Coetzer around 1750, and within twenty years the whole area had been settled. The town of Calitzdorp stands today on a narrow section of alluvial silt soil along the Nel's River, surrounded by higher land that cannot be irrigated. During summer this gives the effect of a long, narrow island of green, ringed by bare brown hills. The central portion of the alluvial belt where the town stands today became a loan farm in the 1750s and was named Buffelsvlei. The loan farm system was gradually phased out after 1813 and Buffelsvlei was granted to the Calitz brothers in 1821. Where the road from Worcester and Ladismith crossed the Nel's River on the Calitz brothers' property the town began to develop. After 1821 subdivision became legal and farms on the river, such as Buffelsvlei, where vineyards and orchards had been planted, were soon broken up into smaller pieces.

The vineyards were heavily irrigated during winter and the moisture obtained by the root systems of the vines was sufficient to ripen large crops of grapes. The soils are alkaline

Ostriches grazed and orchards produced succulent fruits where quality vineyards are now being planted.

Night temperatures are low in the inland area and big falls of snow can be seen on the Swartberg each winter.

and rich. Grapes ripen easily, and even when producing a large crop retain the high levels of acidity required for the production of good brandy. Daniel Nel, who lived further up the Nel's River toward the Swartberg Mountains in the period around 1890, was a famous distiller. The brandy barrels were transported by ox-wagon to Worcester on an annual journey that took between four and five days. They were then railed to Cape Town and shipped to London.

One year, as the wagon with its load creaked along the bumpy Worcester Road, a particular lurch caused a stave in the end of one of the barrels to spring out of place and all the brandy in the cask ran out on to the road. In his anxiety, the wagonmaster, who had noted the area's first telephone line alongside the road, seized the opportunity to warn his employer of his loss. Climbing the nearest pole, he called along the wire, "Oom Danie, Oom Danie, the brandy has spilt!"

Daniel Nel's son, also Daniel, continued the tradition of exporting his farm-distilled brandy to Britain. He was a successful farmer and businessman and eventually left a farm to each of his five sons. In 1890 he bought Die Krans, a farm within the municipal boundary, where he became a major producer of ostrich feathers, sweet wine and brandy.

After the collapse of the ostrich feather trade in 1914 Daniel's sons had to rebuild the family's farming businesses, finding income in other directions. The tradition of commercial brandy distillation was circumscribed by the gradual tightening of control on the distillation of spirits. After the 1920s it became illegal to produce more than 67,5 litres per year.

Carel Nel, Daniel's son, inherited Die Krans, and developed the vineyards on the farm's highly fertile, alluvial soil. He made a little sweet wine, fermented some wine dry to make brandy and dried sultana grapes to make raisins. In the early '30s the raisin market was affected by the Depression, but Carel, like his forebears,

Looking down the Welgeluk vineyards towards the source of the area's summer winds.

was resilient and resourceful. His father and grandfather had been famous brandy distillers, and South Africans were still prepared to pay well for good brandy, no matter how little money was around. A tax was raised on commercial brandy, which was not applied to farm stills, and the Customs and Excise officials used to pay unexpected visits to farm distillers to ensure that the little brandy being made was drunk and not sold.

The big house on Die Krans, where Carel Nel lived, was built by his father Daniel during the years of ostrich feather prosperity. Its wide expanse of ground-level floor space conceals the existence of an extensive below-ground storage cellar. This space became the key to Carel's plans. He started to ferment more of his grape juice dry, and ran the still longer than had been his custom, making more spirit than the permitted maximum. Casks were bought and installed in the cellar and the illicit brandy was stored below ground. On the verandah of the home, Carel built a rack to hold three casks, two of which he filled with brandy and the third with sweet wine. When the Excise inspector came to check, Carel showed him the records of grapes picked, wine fermented and brandy distilled, totalling 67,5 litres, and they sat down to have a drink to break the tedium of an unexciting day. If the Excise man requested a brandy, Carel chose sweet wine for himself. If the inspector preferred sweet wine, the procedure was reversed, and Carel was able to draw attention to his pro-

Boplaas wines are given a period of bottle maturation in an underground cellar.

duction of sweet wine.

Carel Nel normally had more than twenty large casks of brandy in the cellar below his home, the produce of almost his whole crop of grapes. He sold the brandy by mail order and his advertising was done by word of mouth. After an order for a small cask of brandy arrived, together with the required amount of cash, Carel would ride down to the railway station to check the time of the train with the stationmaster. Then he would ring the cellar from the station and ask the foreman to bring a cask of vinegar to the station. The foreman, knowing that vinegar was the code word for brandy, brought the cask up from the cellar, loaded it on the wagon and transported it to the station. Papers were made out for the vinegar, and it was sent off to the consignee.

In later years the KWV offered bottled brandy to grape producers at very attractive prices in return for the surrender of the farm licences. When the Nel family's licence was relinquished, Excise inspectors came and punched holes in the farm's copper still to make sure that it would not produce another drop of brandy.

The little village of Calitzdorp stands on a slight rise beside vineyards on alluvial soil next to the Nel's River.

Two of Carel Nel's three sons, Danie and Chris, joined him on the farms Die Krans and Boplaas, which he had bought in 1936, producing raisins and supplying grapes to the Calitzdorp co-operative. Carel died in 1955 and in 1964 the brothers decided to build a cellar and sell their own wines to the wine trade. The project was successful and in 1965 they bought Welgeluk to increase the supply of grapes to the cellar.

In 1980 the brothers decided to split the enterprise, Chris retaining Die Krans and the Estate cellar and Danie, with his son Carel who had joined the team in 1978, establishing a new wine Estate with Boplaas and Welgeluk. By this time Danie and Carel were committed to the development of a top class cellar producing premium wines. They have established vineyards of Sauvignon blanc, Colombard, Chenin blanc, Cabernet Sauvignon, Merlot, Cabernet franc, Shiraz, Pinotage and Tinta Barocca to replace the varieties planted for bulk production. They have built a modern controlled-fermentation cellar and brought Leon Mostert into the team as cellar master. The plan is to make a range of premium-quality wines — and to draw attention to the potential of Calitzdorp.

The Estate's Boplaas vineyards are planted in fertile alluvial soil where the vines grow and bear grapes with almost unrestrained vigour. By contrast, Welgeluk's red Karoo and sandy loam soils produce smaller crops, with generally better quality grapes.

The Calitzdorp area, like most of the Little Karoo, has a reputation for fierce heat. But Danie Nel points to the south and the source of the Cape's prevailing summer wind. "We're less than 80 kilometres from the sea, to the south-east, and a cool wind starts to blow here after midday most days of the week in summer. We often have to wear jerseys on summer evenings. We don't feel the breeze and the cooler temperatures at Boplaas, where it's sheltered, but Welgeluk faces the south-east and those grapes get lots of ventilation as they ripen."

The mid-season variety of Sauvignon blanc has shown a liking for the Calitzdorp conditions and is producing quality wines in its typically grassy-flavoured style. The Estate also offers a semi-sweet Colombard and a port made from Pinotage.

The grapes for the table wines are picked mid-ripe and are crushed in the Estate's cellar. They are fermented at cool temperatures (15-16 °C) until dry, or until they have reached the required level of sweetness. For the port, fermentation is arrested by the addition of brandy, when about half the sugar has been converted. Both Port and Sauvignon Blanc wines are aged in small casks of French oak.

Estate-made, home-recipe tameletjies (fruit rolls) and a range of jams and condiments are available at Boplaas.

Wine is for sale on the Estate.

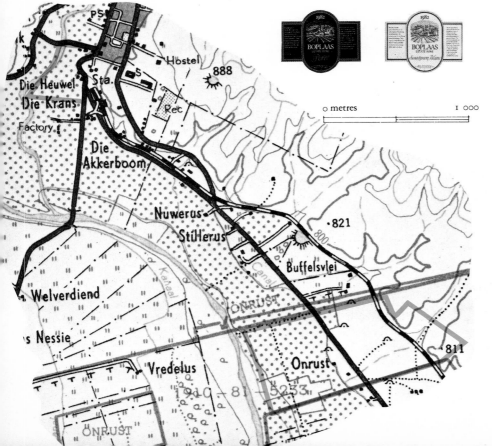

Die Krans

The changing fortunes of the Nel family have followed the fluctuations of prosperity and depression in the Cape's agricultural economy. The Nels of the Calitzdorp area have shown resilience in the face of hardship and audacity in surmounting misfortune. They have created opportunities, taken risks, defied authority, and in some cases hung on to the bitter end.

The first of the individualists in the Calitzdorp area was Louis Nel who took one of the first 5 000 hectare loan farms in the area in 1760. Though Louis was obliged to pay twelve stuivers a year to the Dutch East India Company to stay on the mountainside farm he had chosen, he did not pay any rent for seventeen and a half years. In 1793, the rent for these farms was doubled to twenty-four stuivers and Louis paid nine years of back-rent at the old rate. It is not known if he ever paid the rest of his outstanding amounts, but the Company allowed him to keep the farm. Groenfontein has been divided by inheritance down through the years, but some parts of the farm still belong to his descendants.

Though Groenfontein was primarily a grazing farm, the mountain springs and the Nel's River (named after Louis) provided enough water to be able to grow crops and maintain vineyards. Like other pioneers, the Nels had a brandy still and grape spirit became one of the family's chief sources of income. Grapes grown in the chalky soils that have been eroded out of the Swartberg Mountains have naturally high acidity and are most suitable for the distillation of brandy. The story of the two Daniel Nels, who exported brandy to London, has been told in the previous chapter on the Boplaas Estate.

In 1890 the younger Daniel Nel bought Die Krans, a portion of the original loan farm Buffelsvlei on which the town of Calitzdorp stands today. Die Krans contains 33,5 hectares of vineyard on the edge of town, within the municipal boundary. In addition to brandy, the Nels' income came from the production of sweet fortified wines and the raising of ostriches.

The market for ostrich feathers began to develop from about 1880 and before long most of the farmers in the Klein Karoo had flocks of these birds. The produce of the ostriches, like the Nel brandy, was destined for Europe. Between 1890 and 1914 the major ostrich farmers in the area, including Daniel Nel, became wealthy men. When the World War broke out in Europe, the shipment of ostrich feathers was curtailed and Daniel began to stockpile the produce of his birds. Most of the buildings on Die Krans and his other farms were filled with bags of feathers. When hostilities ceased in 1918, Daniel found that the ostrich feather fashion had vanished along with the old way of life and that his feathers were useless. Though his family had weathered the war years, the feather slump almost lost them the farms and ostriches have not been seen on Die Krans again.

During the period of prosperity, Daniel had bought many farms. Though one had to be sold in 1918, each of his five sons was able to inherit a piece of land on which to rebuild the family fortunes. Carel Nel, who had inherited Die Krans, extended the sultana vineyards and increased his production of sun-dried raisins.

The Calitzdorp railway station is at the end of the line, and the engine has to reverse into the vineyards alongside Die Krans cellar to turn around. On this day, in 1982, a steam engine was used for the last time.

48   Little Karoo

Herds of ostriches once roamed pastures on these alluvial soils. Now vineyards for quality wines are being planted.

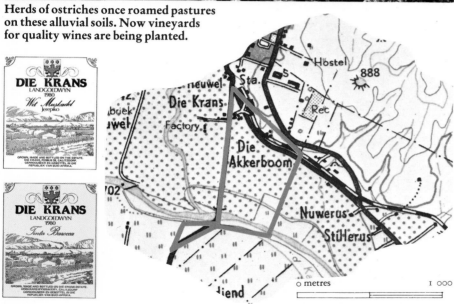

Immediately after the Second World War, two of Carel's sons, Danie and Chris, joined him at Die Krans, farming with vines, drying raisins and making a little sweet wine. After Carel's death in 1955, fluctuations in the income from raisins, the introduction of controlled marketing of dried fruit and a simultaneous increase in the price of fermented table wine caused the brothers to change course and build a wine cellar in 1964. The venture was remarkably successful. They produced low-quality dry white wine and sold it to the eager wine merchant trade. As the production of most Cape cellars began swinging from sweet wine to dry wine, the quality of the wine began to be a factor that influenced its marketability. Danie and Chris planted Chenin blanc, as well as some Tinta Barocca and Pinotage.

During the late '70s Danie's son, Carel, completed his viticultural studies and joined the team at Die Krans, while Chris's son, who is known as Boets, started the same course at Stellenbosch University. During the first flush of wine prosperity Danie and Chris had bought several vineyards around the valley, and in 1980 they decided to divide the operation to allow their sons to have individual enterprises. The Die Krans Estate and several other vineyards and orchards were retained by Chris, and Boplaas, Welgeluk and two other pieces of ground became Danie's property. Danie and Carel built a cellar at Boplaas in 1981 and the Boplaas Estate was born. Boets Nel finished university in 1980 and went into the army to do military service shortly before his father died in mid-1981. He had to obtain six weeks' leave from the army to turn his first harvest into wine in 1982, and the Estate has had to survive a rather tough period.

Sharing a boundary with Boplaas, Die Krans has an almost identical climate. The winters are cold, spring and budbreak are normally fairly early, and the harvest is a week or more ahead of most other quality table wine areas. Though summers are hot, the valley is less than eighty kilometres from the sea to the south-west and the prevailing south-west wind brings some relief on summer afternoons. The average rainfall is about 200 mm per year, falling mostly in March and April, and all the vineyards have to be irrigated.

The soil in the vineyards is a fertile sandy loam, with a very deep layer of topsoil.

Danie and Chris had planted Pinotage, red and white Muscadel, Tinta Barocca, Colombard, Chenin blanc, Hanepoot and Palomino. Boets Nel intends to make top class wines, and time will tell which cultivars will be suited to Calitzdorp's unique weather and ripening cycle. He believes that the farm is best suited to the later ripening varieties and expects demand to be strongest for the red table wines. At present, the Die Krans Estate produces a wide range of dry red and white wines, as well as sherry, port and muscadel wines. The family has been selling Estate wines for several years and has built up a nation-wide list of mail order customers.

Wine is for sale on the Estate.

Die Krans

Doornkraal

This Estate can be found a few kilometres outside the village of De Rust, on the Olifants River forty kilometres east of Oudtshoorn. When it was first settled, in the 1830s, the area known as the Cango was many days' horse-ride away from the nearest pocket of civilisation, at Swellendam. It was a land with fertile soil, little rain, irregular river water and a great deal of hardship.

There is a house standing in the hills above the village, built by one of Gerrit le Roux's forebears in 1850. On the first day of construction, the four corner poles of one of the rooms and rafters for the ceiling were erected and made sturdy before nightfall. That night the builders slept on a platform on the rafters. Lions roamed the area at night and the pioneers had to be inventive to stay alive.

Most of the difficulties of life in those days were connected to the scarcity of water. The farmers could not depend on any crop without irrigation and constructed furrows that connected upstream with the river and led by gravity to the planted fields. But the flow of the river was irregular and often brutal. The Olifants River has more than 3 000 square kilometres of catchment area, but all within the Great Karoo, where the average rainfall is less than 100 mm per year.

When storms break in the Great Karoo, the water washes quickly off the hardbaked ground, creating a flash flood that courses down the dry beds of rivers on its way to the sea. If the interior has substantial rain, there will be enough water in the river for three or four days' flood irrigation of the crops and vineyards. The families work day and night diverting water, until the level has dropped below the entrance to their furrow, and the flood has passed on down to the ocean.

P. M. le Roux, great-grandfather of Gerrit le Roux, the present owner, was born on his family's farm just outside Oudtshoorn. He sold his share of the family inheritance to his brother and moved to the Rietvlei area, forty kilometres away. P. M. le Roux grew fruit, grapes and tobacco, most of which were dried. Grape juice was fermented into wine, then distilled into brandy. He traded by bartering his goods for sheep and cattle, which were as good as money. He had thirteen children and used his sons as salesmen, heading them north and east into the interior to barter with dried fruit, tobacco and brandy. Each of his children was left a farm in his will.

One of P.M.'s sons, Gert, inherited the farm at De Rust, known as Doornkraal, and like his father became an extremely successful farmer, this time with ostriches. Oudtshoorn is ideally suited to ostrich raising and Gert le Roux concentrated on making money during the golden days of the ostrich boom. He paid less attention to tobacco and agricultural crops, and grew only lucerne for the ostriches and a small vineyard for sweet wine and brandy production. Grazing cattle and raising ostriches were his chief interests.

The money made from the ostriches paid for a magnificent house and a revolutionary pump powered by gas-engine to extract water from the swollen river more efficiently. Such was the scale of this purchase that it was despatched from Scotland with an engineer to supervise its many loadings and off-loadings, to install it at De Rust and to teach its new owner how to work it. This engine contained a furnace, in which coke was burnt, producing gas to drive the pump. Gert le Roux hired the Scottish engineer to live on the farm and to make himself available to start the pump the moment the river began to run, which occurred three or four times a year. When the price of ostrich feathers fell in 1917, Le Roux's farming enterprise took such a blow that it could no longer support the Scotsman. Gert's son was taught how to start the engine and he carried out this duty for almost twenty years. In 1942 he was elected

Gerrit le Roux, leading producer of dessert wines, and his son Piet, who will make dry table wines.

Between De Rust and the Stompdrift Dam, Doornkraal's rich soils and cool temperatures produce top dessert wines.

Klein Karoo

to Parliament and each year had to move to Cape Town when the House was in session. When the rain started falling in the Great Karoo he would get a telephone call, upon which he would make a rushed car trip of five or six hours to Doornkraal to start up the pump. He then would return to the business of the nation in Cape Town. Eventually one of the farm workers took over this duty, and finally in 1946 the pump was fitted with a diesel engine.

In 1927 a dam was built on the Kamanassie River, a tributary of the Olifants River, and in 1969 the Stompdrift Dam was built on the Olifants. These dams have largely controlled the irrigation supply of water, and more intensive year-round agriculture is now practised.

Gert le Roux had a very small cellar where he made sweet wine and distilled brandy for his family and friends. A glass of dessert wine was part of the culture of those times. When the ostrich bubble could be seen to have burst, Gert called his son P.K. home to take charge of the farm. A new approach was needed and a new source of income had to be found. P.K. began to extend the area under vineyards chiefly with red and white Muscadel. He was unable to sell any of his product as good wine that first year. But resisting the economic pressure he matured it in the cellar. The following year, merchants were interested in the older wine, but did not want the product of the new harvest. P.K. held out, found a sympathetic merchant and finally became a leader in the sweet wine industry.

Gerrit le Roux studied viticulture at Stellenbosch University and took over the running of the farm in 1959 from his father, who had combined this duty with his parliamentary role. He has further expanded the vineyards, planting Steen, Colombard, Sémillon, Tinta Barocca and Pinotage. All are used for the production of fortified wines.

Doornkraal has rich alluvial soils beside the river and a mixture of enon conglomerate and Table Mountain sandstone on higher ground. The less fertile soils have proved to produce better quality wines. Less than sixty kilometres from the sea, Doornkraal has cool evenings in summer and grapes ripen in these vineyards comparatively late.

"We obtain high sugar levels in our grapes quite easily here," says Gerrit, "and I have no doubt that the more highly-regarded varieties used for making dry table wines will do well. We still have to do evaluations, but I believe that the Little Karoo area will follow the Breede River Valley in moving from purely sweet wine production to more dry wines." He makes red and white Muscadel, sweet Steen and port-type wines of distinction. In the last major area producing large volumes of sweet wines, he has often won the prize for Champion Wine.

Most of the wines are sold to a merchant, but a small quantity is retained for sale on the Estate. It is expected that this quantity will increase and that Gerrit and his son will also make dry table wines in the next few years.

Wine is for sale on the Estate.

The ostrich boom (1880-1914) brought great prosperity to Oudtshoorn. Gert le Roux built the Doornkraal home in 1914.

52   *Paarl*

Backsberg

In recent years, some wine Estates have drawn particular attention to themselves because of innovation and consistent quality within a range of wines. Among the foremost of this group is the mountainous Paarl farm known since 1969 as Backsberg.

The farm has a number of natural advantages, but the growing conditions are not dissimilar to those on many other Estates. As Sydney Back says, "Vines are comfortable here." Yet one has to look at how the human element shapes the conditions under which the vines grow and produce their grapes to find the factors that are contributing to Backsberg's success. Sydney and Michael Back, a father and son team, have their attention firmly centred on the vineyards and believe that most of the improvements in the quality of their wines can be traced back to a correct decision made in the growing of the vines.

Backsberg is part of a freehold grant made to Pieter van der Byl in 1692 and subsequently enlarged before subdivision. Van der Byl, who was aged eight when he arrived in the Cape with his parents in 1668, named this farm Babylons Toren after the curiously-shaped mound of rock that stands in the valley north of Simonsberg Mountain.

Opponents of W. A. van der Stel's high-handed form of administration, Van der Byl and his friends Hüsing of Meerlust, Appel of Vergenoegd, Meerland of Meerendal and Van der Heyden of Klein Welmoed were arrested and deported in disgrace to Holland in 1706. Unfortunately for Van der Stel, the revolutionaries' arguments were heard in Holland and the Cape administration was placed under investigation by the head office of the Company. W. A. van der Stel and his senior officers later resigned. The following year Pieter van der Byl returned to Babylons Toren with his young wife. After Hester's death, the farm was sold to Johannes Louw.

In 1813 and 1819 the farm was enlarged with quitrent grants, most of which were grazing ground, as they included the lower and higher slopes of the northern side of Kanonkop Hill. A 28 hectare section of adjoining land was granted to C. E. Ponty in 1813 and this was later amalgamated with Babylons Toren. In 1825 Ponty had 180 000 vines and 140 leaguers of wine on Babylons Toren. In 1916, a 220 hectare portion of Babylons Toren was sold to C. L. Back, who had arrived in the Cape from Lithuania in 1902. C. L. first supported himself as a butcher boy, earned extra money by working on the breakwater in Cape Town harbour, and finally found enough money to buy a small butcher shop in Suider Paarl. After further saving, he sold the butchery and set out to build a new life on his farm on the slopes of the mountain.

Klein Babylons Toren, as the farm was known, had a small vineyard in the shallow kloof on the lower part of the farm, apple and orange orchards, and some grain fields. Most of the property was covered with bush and pine trees. C. L. cleared most of the lower sections of bush for vineyards, apricot orchards, and pastures for sheep and dairy cattle. Running a mixed farm was a common practice at that time. He grew Cinsaut and Sémillon vines and made sweet wines that he sold to the KWV. From 1930, he began to export dry red wines made from Cinsaut and Shiraz in oak casks to Canada and Britain.

The vineyards consisted of bush vines and, not being irrigated, produced small crops. These low-yielding vines were able to ripen their grapes to high sugar levels and C. L.

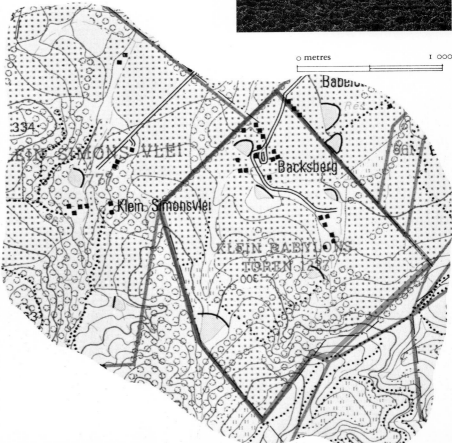

used them to make fine quality sweet wines, winning the Burgoyne trophy at the National Young Wine Show in Cape Town in the mid '30s. Sydney Back joined his father at Klein Babylons Toren in 1938 and assisted in the production of the dry red and sweet wines for export. In 1939, Britain's entry into hostilities with Germany made the shipping of wine impossible.

Sydney proceeded to remove bush and prepare the soil for more vineyards to increase the farm's production of wine. He continued to supply his wines to the KWV after the war, and in 1950 applied for a wholesaler's liquor licence, which would enable him to increase production and sell wine to a wide range of customers. He installed bottling and packing facilities at the cellar he now called Back's Wines, and started building a business supplying bottle stores with low-priced 'canteen wines', as they were called. Sydney expanded the production of the farm's vineyards, but the business grew at such a pace that he had to buy wines from co-operative cellars and grapes from surrounding farmers to supply the demand. In the mid '60s, he planted Cabernet Sauvignon and decided that the production of varieties with low yields but high returns would be an answer to his problem.

In 1969 another wholesaler, restricted by a small quota, bought the Back's Wines wholesale business, thus allowing Sydney the capital and freedom to venture into the production of Estate wines. Sydney renamed his property Backsberg, and planted Cabernet Sauvignon, Pinotage and Shiraz.

Sydney Back, like most other leading wine producers, makes the type of wines that he personally enjoys most. Because he preferred to drink red wines of a lighter body than was currently being produced in South Africa, he planned his vineyards and harvesting procedure to make fresher, fruitier wines that matured earlier than the heavier styles. Because he preferred to drink fuller white wines, in the style of the whites of Burgundy and the Loire, he established this style as the objective of Backsberg. He planted Chenin blanc and Cape Riesling vines to boost the quality of the white wines. In 1972, he was able to obtain some Chardonnay and Sauvignon blanc vines and began to propagate these to be able to provide enough planting material. In addition, he planted vineyards of Bukettraube and Kerner. Recently, Sydney has added Merlot and Pinot noir vineyards to the red varieties and Weisser Riesling to the whites.

A third generation entered the picture when Michael Back, a graduate of Stellenbosch University, joined his father and John Martin on the farm in

The granitic outcrop called Babylons Toren by the first settler, is situated on the neighbouring property.

Backsberg's cellar complex is at the foot of the Kanonkop slopes, with generally white vineyards higher and red vineyards lower.

The premium white varieties have been planted on the cooler growing conditions of Backsberg's higher slopes.

1976. John Martin has been a key figure in the Backsberg administrative team since the mid '50s, when Back's Wines was beginning to grow, and is in charge of sales and marketing.

Backsberg has approximately 160 hectares of vineyards, extending from a small area of relatively level land in front of the cellar, across the twisting slopes of the lower Kanonkop and up terraced ridges to a point 250 metres above the cellar. Like most Cape wine producers, Sydney planted his first Cabernet Sauvignon vines in what were then the farm's higher vineyards. He has subsequently learnt however, that in these cool situations, on high-potential soil, this vigorous vine puts too much energy into growth and not enough into its crop. He believes that the quality of the production from Cabernet Sauvignon is superior when grown in the sandier soils lower on the farm. The Estate has roughly 25 per cent of sandy loam soils, which have a base layer of clay for moisture retention, and the rest is reddish, mountain soil, produced by decomposed granite. This soil is deep, but has a degree of moisture retention to help summer ripening.

The recently cleared and terraced slopes at the top of the farm and most of the intermediate slopes are planted with early and mid-season white varieties. The average rainfall in this area has been measured to be, over an eleven-year period, 800 mm per year, and this above-average rainfall is supplemented by an irrigation system installed in 80 per cent of the vineyards. The addition of water is restricted to periods of stress, before the ripening season or after harvesting. The farm receives some benefit from the south-east wind during summer, but is sheltered from the more severe effects of this damaging force.

To make the preferred lighter reds, Backsberg's grapes are picked at full-ripe stage and, after crushing the mash of skins and juice is fermented at controlled temperatures in closed tanks until the required degree of colour has been obtained. After removal of the skins, the must is fermented dry. Once the malolactic fermentation is done, the Cabernet Sauvignon and Shiraz wines are placed in large oak vats for about one year's maturation. The Pinotage remains in steel until sufficiently mature to bottle. Pinot Noir is scheduled for maturation in small oak casks.

The white wines are made with several hours of skin contact and are fermented in closed stainless steel tanks at cool (15 °C) temperatures. After fermentation, the dry wines are given a period of maturation, with Chardonnay and Sauvignon Blanc going into small oak casks and the rest remaining in steel. Semi-sweet wines are made by arresting the fermentation of fuller, flowery wines at the required degree of sweetness. These wines remain in steel until bottling time.

A major feature of Backsberg's range of wines is the availability of matured vintages of its wines. Backsberg usually has five vintages of Cabernet Sauvignon, three of Shiraz and two of Chardonnay and Sauvignon Blanc available at the one time. Often demand outstrips supply.

Wine is for sale on the Estate.

Boschendal

Production of top white and red wines commenced at Boschendal in 1978.

At the entrance to the Estate, the manor house, restaurant and gift shop provide a welcome for the visitor.

The Boschendal manor house has been restored as an eighteenth-century museum.

"We've chosen to go the quality wine route. We've planted quality cultivars. We've got a very modern cellar and we've got a fantastic team functioning here to produce the very best wine. It's a hell of a lot of work. You need determination. You just have to go and get it." These are the words of Achim von Arnim, extroverted cellar master and individualist who has been linked with this Estate since 1978.

Boschendal is the largest wine Estate in South Africa, having vineyards situated on various portions of the conjoined Rhodes Fruit Farms, which cover 3 000 hectares of land between the eastern slopes of the Simonsberg and the Berg River as it courses through the Franschhoek Valley. Today the amalgamated property contains seventeen originally separate farms, and the distance between the most eastern and western wine grape vineyards is 13,5 kilometres. The land within the boundary supports, in addition to vineyards supplying the Estate cellar, table grape vineyards, orchards with plum, peach, pear, apricot and lemon trees, commercial forest, a nature reserve, a cannery and several hundred staff.

The Boschendal manor house and surrounding land played an important part in the settler heritage of South Africa. Boschendal was owned by members of the De Villiers family for a period of 169 years and was considered to be the ancestral seat of the family for most of this time. The land was originally promised to Jean le Long, a land surveyor, by Governor Simon van der Stel in 1685. Le Long was one of the first French Huguenots to take advantage of the Dutch East India Company's offer of refuge at the Cape. He arrived with his Dutch wife, Daniella, and took possession of the 51 hectare farm in the forest at Groot Drakenstein.

In 1713 Le Long was officially granted Bossendal (the original spelling) on condition that he would build a road beside his wheat field, leading to the ford across the river. The road was to be for his own use, as well as for the other inhabitants of the valley. He was to give one tenth of his annual grain harvest to the Company and he was to replace all trees felled for any purpose with oaks or similar trees.

On a similar-sized plot, next to Bossendal, lived Le Long's friend, Nicholas de la Noy. De la Noy had been given permission to work this land, which he named Champagne, in 1690. In 1692, when the first Cape census was taken, De la Noy was alone on the farm. He had 1 500 vines, 1 gun, 1 heifer, 6 oxen, 2 pigs and a cow.

In 1715 Abraham de Villiers bought Bossendal and Champagne, and amalgamated the properties under the name of Boschendal. Abraham was one of three Huguenot brothers who reached the Cape from Holland in 1689. They had left the rest of their family at Niort, near La Rochelle in France, and fled to the Netherlands in 1685. From there they were given free passage on the ship *Zion* by the Company, which also provided them with a letter to Simon van der Stel, praising their knowledge of viticulture. According to his custom, Van der Stel promised them land where they could demonstrate their skills and determination as farmers. Before setting out for the valley under the Groot Drakenstein mountain, they were supplied with provisions, farming implements, seeds and planks to erect a shelter.

The brothers had permission to work on a communal farm of 51 hectares, which they named La Rochelle, but which was never officially granted to them. Their performance as farmers evidently satisfied Van der Stel, as each was granted adjoining similar-sized farms elsewhere in the valley in 1694. Pierre, the eldest, named his farm Bourgogne, Abraham his Champagne, and Jacques, the youngest, his La Brie. Abraham de Villiers sold both of his Boschendal farms to his brother Jacques in 1717. In the census of that year, Abraham declared his ownership of 30 000 vines and 13 leaguers (7 500 litres) of wine.

After the death of Jacques de Villiers in 1735 Boschendal was run by his wife Marguerite and their son Jan (Jean) until the latter's marriage in 1738, when he became sole owner. Jan had twenty-two children and is believed to have built the forerunner of the Boschendal manor house. After he died in 1796 his second wife, Gertruida, farmed Boschendal until her death in 1807, when it was inherited by her son Paul.

According to the De Villiers family the original date on the Boschendal gable was 1746, indicating that it had been built during Jan's period of ownership. In 1812 Paul rebuilt the already ancient home in the shape it may be seen today, and placed the year, together with his and his wife's initials, on the gable. He sold half of Boschendal to each of his two sons, Jan Jacobus and Hendrik, in 1840.

Jan Jacobus sold his share to his brother in 1843, then in 1860 bought back both pieces of the farm before selling it out of the family in 1879, after his son, already farming on Harmony, had declined the family's offer to purchase Boschendal. Eighteen years later Boschendal was bought by Cecil John Rhodes as part of his plan to buy the major part of the Groot Drakenstein valley and turn this area into the deciduous fruit growing centre of the Cape. Within a short period in 1897 Rhodes's agent, Lewis Michell, had bought ten farms for £30 000. Before long there were twenty-nine farms in Rhodes's ownership, being operated by twelve managers, all trained in California. At a meeting in March 1902 Rhodes, the

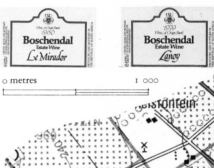

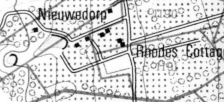

Boschendal's Riesling is picked early to make a light drinkable style with fresh, fruity flavour.

De Beers Diamond Company and Alfred Beit formed Rhodes Fruit Farms to operate the properties and build a fruit-growing and canning empire. Unfortunately, Rhodes died the next day.

De Beers sold their interest in RFF in 1936, but in 1969 they agreed, together with the Anglo American Corporation and the Rand Selection Corporation, to buy the majority shareholding in RFF again. During the intervening years RFF had led South African agriculture in the production and processing of deciduous fruit for export to Europe. The farm produced table grapes for export and maintained a cellar and a few vineyards for sherry production. The sherries were made by Ronnie van Rooyen, who worked on the farm for many decades. At that time, there was a much larger market for sherry than table wine in South Africa, as the techniques and equipment necessary to make clean dry wines were not yet available.

The boundaries of both RFF and the Boschendal wine Estate, covering seventeen original properties, contain very many differing soil types and growing conditions for both vineyards and orchards. There are three main soil types. High under the cliffs of the Simonsberg clay-loam slopes provide sheltered growing conditions. The central part of the property, where most of the vineyards are found, has mixed Hutton and decomposed sandstone soils. Down by the river, some 260 metres below the most elevated vineyards on the Simonsberg, there is a mixture of sandy alluvium and sediment mixed with small sandstone boulders, providing efficient drainage.

"We have restrictions on this farm that may be seen as limitations or as advantages, depending on your point of view," says Achim von Arnim, the Estate cellar master. "We don't make the wine. The farm grows it. That's why Boschendal wines will always be delicate and fruity. The conditions were always here. We just help the wine into the bottle."

Achim is the spokesman for a dedicated team that have lifted Boschendal's image into the leading group of Cape Estates. He pays tribute to Herman Hanekom, vineyard chief, Mynhardt Theron, quality controller and Erasmus Pretorius, cellar technician, for their contribution to the team that cares for the Estate vines and wines.

The Boschendal Estate runs up the eastern slope of the Simonsberg from the Berg River and faces the sun rising over the Franschhoek mountains. The sun sets over the peak of the Simonsberg comparatively early in the afternoon and gives Boschendal at least an hour's less afternoon sunshine than west-facing farms. "We believe that the morning sun is better for our vines than the afternoon sun," says Achim. "Our vineyards are wet with dew when those first rays of the sun arrive. And they cool down gradually during an extended dusk. These conditions do not produce high sugars, but we have found that to be an advantage. We have planted Sauvignon blanc, Gewürztraminer, Weisser Riesling, Pinot noir, Sémillon and Merlot, all early to mid-season ripening, and all of them classic quality cultivars."

Because of the geographical layout

Boschendal uses the best available equipment to maintain the quality in the grapes throughout the cellar process.

The elevation and east-facing aspect has influenced Boschendal to plant early-ripening cultivars.

of the farm, Boschendal has grapes down in the valley that are fully ripe at 19° to 20° Balling, and at the same time others higher up the slopes are ideally ripe at 22° and 23° Balling. "That produces a natural range of wines," says Achim, "wines that are actually different by their birth.

"We keep a lot of records on the performances of our vineyards. We know, for instance, that when our sugars on the deep alluvial soil at the bottom of the farm come to 19,4 and they've got 7,3 acidity and a pH of 3,4, then we're very close to picking. We can never expect that vineyard, from its history, ever to produce more sugar without a loss of acidity in turn. But we know that in our Nuwedorp vineyard we can go up to 23° Balling without losing acidity. To produce a delicate wine, you have to handle it like bone china. Any mistake we make with these delicate wines will immediately make them either thin or they'll get a knock and they'll never arrive in the bottle as quality. So we have to first pick them with sufficient acidity, which is a natural protection. We try to keep a reasonable pH in the grapes, as well as in the juice and finally in the wine. Then we take not only the free-run juice from our drainers, but we add the pressed juice from our tank press to give it some extra backbone."

Boschendal has 250 hectares of vineyards, producing more than 2 000 tonnes of grapes. The harvesting of these grapes is a major task involving vineyard planning throughout the year, a highly organised picking programme and speed in crushing the grapes and storing the chilled juice. A major contribution to the quality of the Boschendal range of wines may be attributed to the efforts of the team in the vineyards. Herman Hanekom was originally brought to Boschendal to look after the vineyards of table grapes. His arrival roughly coincided with the Estate's developing interest in the production of quality table wine. Herman's natural enthusiasm doubled the size of his job and the wine vineyards today take up the majority of his time and interest.

Approximately 75 per cent of Boschendal's vineyards are planted with white cultivars. Historically, most of these were planted with Steen vines, and this is the reason why the Estate has a number of Steen and Steen-based wines in the range. Even before the addition of the extra flavours and styles of the newer cultivars, the disparity of character provided by the different Steens from different parts of the farm was the basis of a range.

"We only take 700 litres of quality juice off a tonne of white grapes," says Achim. "The rest goes to distilling. From the 700 litres of cloudy juice, we get 630 litres of clear juice that goes into fermentation. We inoculate this juice with a selected culture of yeast and ferment it from 14° to 16°C. We don't have robust juice that can handle really low temperatures. Our clear, delicate juice needs more yeast activity than heavier juice, and the yeast definitely imparts a character to the wine. We have a fermentation period of two to three weeks. We then clean up the wines and get them ready for blending.

"We initially keep our wines separate, each vineyard in a different tank. So we have the whole scale of wines, from dry to semi-sweet, running up the geographical scale of the farm, in tanks as they were picked, with one difference. We interrupt the fermentation of some of our fuller wines, so that they provide their own sweetness. This is a great advantage. The same grape, coming from the same soil, from the same fermentation, having its natural sweetness.

"We've evolved a method where we interrupt the fermentation, fine the wine, filter it and only then give it a little sulphur. That's how we protect our semi-sweet wines. We keep these at –4 °C in the stabilising tank and bottle them as soon as possible."

Cabernet Sauvignon grapes are used for the base of the light, fruity blended Lanoy.

To protect the health of the vineyard, the spraying programme begins in early summer.

Facing east, the vineyards receive the first rays of the early morning sun.

Boschendal red vineyards are planned to provide 25 per cent of the farm's wine production. Varieties planted include Cabernet Sauvignon, Shiraz, Pinot noir, Tinta Barocca, Merlot and Pinotage. Once again the emphasis is on earlier-ripening cultivars, with the exception of Cabernet Sauvignon, a late variety and a difficult customer to place. Cabernet Sauvignon vineyards planted on different parts of the farm have provided a number of different styles of wine to date and have helped extend the range of products offered by the Estate. Lanoy, Boschendal's premier red blend, shows that the farm is capable of producing a highly aromatic red with distinct Cabernet characteristics. Blanc de Noir, a white wine produced from red varieties planted in areas that restrict their ability to ripen, has caught the interest of the public.

"Our operation has to be based on reality and practicality. Boschendal vineyards produce delicate wines and that goes for the reds too. So we accept this situation and we treat our red wines in such a way as to make the most of the advantage that nature has given us," says Achim.

"We pick at from 21 ° to 23 ° Balling. We ferment the juice on the skins in a closed vessel, where we hold the temperature at between 18 ° and 20 °C, until 6 ° to 8 ° Balling. Then we press all the liquid out of the skins and ferment it dry. We mature our red wines in large vats of Nevers oak. I think our more delicate reds mature in these larger vessels more positively. We now have enough wood to mature two vintages of red wines. We need that quantity because we're giving the wines from twelve to eighteen months in wood. The wines are then bottled and given at least another year of bottle maturation."

In 1980 Jean-Louis Denois, a young Frenchman from Épernay, brought another talent to Boschendal and provided the impetus for the introduction of a bottle-fermented sparkling wine. Jean-Louis and Achim von Arnim instituted a programme aimed to produce low-alcohol, high-acidity base wines that would be fermented into a high quality sparkling wine. Achim believes that Boschendal's Cape Riesling provides the necessary neutrality required by the base wine. Each of the handling processes that follow are applications of those used in Champagne. The wine has almost twelve months' maturation on the lees within the bottle, and it provides the classic refreshment that is the trademark of Champagne and Méthode Champenoise.

"Can we produce 2 000 tonnes of quality wines?" asks Achim. "If we have the quality cultivars, the equipment to handle them with the necessary speed and a motivated team, using the best system we can think of, why not? We'll certainly give it a try."

The old Boschendal cellar beside the manor house has been converted into a restaurant that serves traditional food of the Cape. Another historic building offers canned and bottled produce of the farm orchards and garden.

Wine is for sale on the Estate.

Boschendal's high slopes, mid slopes and river flats provide a natural range of wines.

De Zoete Inval

With large farms and a long harvesting season, South Africa has few wine producers that rely on the grapes of a handful of varieties to make their wines. Most have ten or more varieties, to stretch out the season. Adrian Frater of De Zoete Inval has 50 hectares of Cabernet Sauvignon vineyards, and this one variety produces 90 per cent of the wine made in the Estate cellar, with the rest coming from small blocks of Pinot noir, Cinsaut and Chenin blanc. The Cabernet Sauvignon and 13 hectares of table grapes provide the farm with most of its income.

Adrian Frater has specialised in Cabernet Sauvignon because this classic variety is best suited to the difficult conditions on his farm. De Zoete Inval is adjacent to the town of Paarl, on the banks of the Berg River, just above the junction with the Van Wyk's River. It has broad flat areas of alluvial, mostly sandy soils, which have low moisture-retention qualities. To contrast with these extremely deep soils, other areas of the farm have drainage problems, where water accumulates and, until a drainage system was installed, prevented successful use of the land. The Frater family have installed 50 kilometres of pipes, from 1,5 to 3 metres below the surface, to help water drain away after winter.

Cabernet Sauvignon is grown in most of the De Zoete Inval's vineyards because it has proved to thrive in the deep soils. Cabernet is a vigorous plant and is able to survive difficult conditions. The table grape vineyards receive additional water in the form of sprinkler and drip irrigation, and the Cabernet Sauvignon vines receive assistance from overhead sprinklers in summer. The climate in the very base of the valley, beside the river, is hot in mid-summer, with a cooling breeze coming from the south-west in the late afternoon.

This spot, beside the Berg River, was first chosen by Hercules des Prez, who came to the Cape in response to the offer of land and opportunity to the French Huguenot refugees. Though born in France in 1645, Des Prez was recorded living in Courtrai in what is now Belgium between 1668 and 1678. In 1686 he joined the Sint-Jans guild of tailors in Flushing, in what is now the Netherlands, and two years later accepted the Dutch East India Company's offer to emigrate to the Cape.

Hercules voyaged to the Cape with his wife Cecilia and their six children on the 42-metre sailing vessel *Schelde*. The Des Prez family and their seventeen refugee companions had an exciting trip: seven or eight days out at sea, a terrible storm developed and the ship was forced into Porto Prayo, on St. Iago. On arrival, the passengers learnt that an English pirate vessel had captured three ships belonging to the Dutch, English and Portuguese, the previous day, just off the coast. The *Schelde* took to sea immediately and the rest of the voyage was uneventful until the ship ran into another storm, just short of the Cape.

Hercules settled his family on the banks of the Berg River, not far from the Taillefert family and the Le Roux brothers beside Paarl Mountain. Having lived in Belgium and Holland for some time, he spoke Dutch and gave his new farm a Dutch name.

In 1690 he applied for and was granted assistance by the Company to feed his family. He died around 1695, and the farm was officially granted to the family in 1698. The eldest son, Hercules Jnr., was an active conspirator against W. A. van der Stel and a leader of the cause of burgher rights. The farm was sold by the Des Prez family in 1728 and passed through the hands of five more owners in the next ten years. The riverside property was found to be prone to winter and spring floods, which were only alleviated in later years by the widening

On the eastern border of the town of Paarl, De Zoete Inval lies in the floor of the Berg River Valley.

Situated just above the junction of two rivers, De Zoete Inval has mostly sandy soils with low fertility.

of the river bed and the construction of a dam at Wemmershoek to supply water to Cape Town.

In 1880 De Zoete Inval was bought by the firm Mossop and Frater to provide soft water for the washing of wool. Robert Frater had owned a wool mill on the English-Scottish border, and had migrated to South Africa when the mill burnt down. His wife had died in childbirth shortly beforehand, and Robert left his surviving infant son, John Robert, with his sister. He came to Kimberley and worked on the diamond fields. Unable to find his fortune on the diggings, he came to Cape Town and sought employment in the wool trade. With Mossop, he founded a wool washery at De Zoete Inval. When Robert Frater's sister died, John Robert was fifteen years old, and his father brought him to South Africa to discover a new land. In time, the wool washing business declined and Robert began to develop the farm and built a small cellar for the produce of his vines.

After John Robert took over the farm, he sold his wines to the KWV. Some ports made from Cinsaut and Portuguese varieties were exported to London. He was succeeded by his son Gerard, who continued to supply the KWV with sweet wines. Adrian Frater, John Robert's grandson and the present Estate vintner, made a port wine in 1955 that won the General Smuts trophy for champion wine.

When Adrian saw how well Cabernet Sauvignon adapted to the farm's conditions, he extended the vineyards of this cultivar. He harvests the Cabernet at mid-ripe stage (21 to 22 ° Balling) and ferments the mash of skins and juice in open concrete tanks, with a limited amount of cooling, until sufficient colour has been extracted (usually at between 10 and 15 ° Balling), and ferments the must dry in closed fibre-glass tanks. The wine is racked and matured in the same tanks for three years before bottling and bottle maturation. In general, each vintage of Cabernet Sauvignon is released when the preceding vintage has sold out. Occasionally, small batches of slow maturing wine are held back.

The Cinsaut, Pinot noir and Chenin blanc vineyards produce cultivar wines. The grapes of these varieties reach a higher sugar content than those of Cabernet Sauvignon, and make fuller wines with lower acidity. Adrian Frater markets the Chenin Blanc wine under the name Capri. The Pinot Noir is simply described as Extra.

Wine is not for sale on the Estate.

The deep-draining soils present the vines with moisture problems in summer. Supplementary sprinkler irrigation is used.

Most of De Zoete Inval's wine-producing vineyards contain Cabernet Sauvignon, best suited to the arduous conditions.

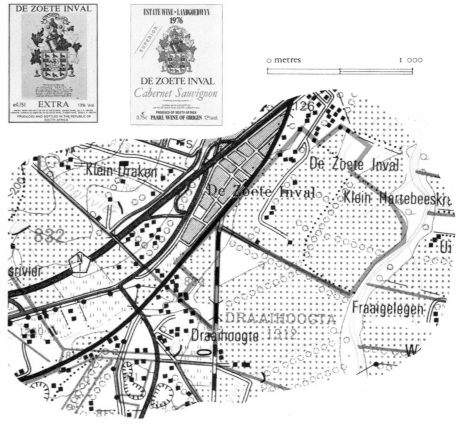

De Zoete Inval

Fairview

The Fairview cellar stands at the foot of Paarl Mountain. Most of the vineyards are further down the slope.

Long recognised as one of South Africa's premier red wine producing farms, Fairview was for many years a major supplier of full-bodied red wines to the KWV for their export programme. During the early 1970s, when South African wine Estates began to sell significant quantities of their wine to an interested consuming public, Cyril Back made preparations to convert Fairview into a direct marketing operation. The first opportunity the South African public had of tasting Fairview Estate wines was in 1974, when the Back family hosted the first Estate auction in the country. Since then, Fairview Estate wines have been available each year from the Estate.

Fairview is known to have once been part of the Blomkolsfontein farm, which extended over a great part of Paarl Mountain. In the early 1900s Blomkolsfontein was owned by a Mr. Hugo. He sold the highest part of the farm and built the Fairview home on the remainder in 1917. In 1922 he built the cellar on Fairview and made wine for sale to the public. Most of the vineyards on the farm at that time are believed to have been Cinsaut and from them Mr. Hugo made dry red wine, which he sold in small barrels to both the public and the trade.

Mr. Hugo died in 1937 and the farm was bought from his estate by Mr. C. L. Back, Cyril's father, who had previously run a wine-farming business on what is now known as Backsberg (owned by Cyril's brother Sydney). One of C. L. Back's first projects on Fairview was to replace most of the old Cinsaut vineyards with Steen and Pinotage. He had long supplied the KWV with dry red wine, and white and red port-type wines from the farm on the slopes of Simonsberg, and he continued to do so from Fairview. On his death in 1954, Cyril Back took over the reins of Fairview and continued in the traditions established by his father. He replanted with quality red cultivars to the extent that there were no Cinsaut vines left on the farm.

"This farm is ideally suited to the production of full-bodied wines," says Cyril Back. "None of the soils are sufficiently fertile to give large production from any cultivar. I believe that quality comes first. The first red varieties that were established on this Estate were Cabernet Sauvignon, Shiraz, Pinotage and Tinta Barocca. The full-bodied wines made from these cultivars won many prizes at wine shows and reassured us that the public wanted reds with plenty of flavour and character." Cyril made cultivar wines that built the Fairview reputation from Cabernet Sauvignon, Shiraz and Pinotage. In addition he made fortified white dessert wines from Steen and Sémillon vines.

In 1978, Cyril was joined in the Fairview cellar by his son Charles, and

South Africa's first private Estate auction in 1974 indicated the nation's interest in Fairview.

Looking across the Paarl Valley to the south-east, toward the gap between Klapmutskop and Kanonkop.

In very early spring, the vines flower and the bunches are formed.

several significant developments have followed the change.

"These farms along the Paarl Mountain were famous for reds, and didn't have cellar facilities for white wine making," says Charles. "But 35 per cent of our vineyards were planted to Steen and Sémillon. I wasn't interested in making ports, so we installed the first of our stainless steel."

Since that date, all of the old concrete fermenting tanks have been removed and replaced by stainless steel. The 1922 cellar has had two major refits and is now equipped to ferment all of the Estate wines dry under closed, temperature-controlled conditions.

"We get high sugar here very easily," says Charles. "The vineyards produce rich, sweet grapes and they, in turn, make full wines. So I don't intend to change the style."

One change that can be seen in the red wines may be a result of fermenting at lower temperatures in closed containers.

"The cooler you ferment, the less tannin you drain out of the skins, and the longer you can ferment on the skins, the more flavour you get. You also make a more aromatic wine, a better balanced wine."

Charles also sets out to make full-bodied whites. "The farm makes the grapes, I just turn them into wine. We get high sugar with sufficient acid in almost all our vineyards. So our dry whites will have plenty of body and flavour. The varieties that are being used for these heavier whites are Steen, Sauvignon blanc, Sémillon and Bukettraube."

Steen has been used to make full and hearty dry and semi-sweet wines for the past few seasons. The Sémillon has, until recently, continued to produce a fortified wine. Charles plans to make blended dry whites from these two varieties in conjunction with Sauvignon blanc, which has adapted well to the soil and growing conditions.

"I like to have the white grapes arrive in the cellar at 23° to 24° Balling. And I try to get as much contact between the skins and the juice after crushing as is practically possible. So we've installed de-juicers that allow the crushed skins and juice to stand before we drain off the juice for fermentation. From there the process is standard. I'm aiming to make full-bodied, extra-dry, wood-matured white wines."

Because of the ease with which the Estate's white cultivar vineyards develop high levels of sweetness, Fairview is ideally equipped to produce richly flavoured Late Harvest and other semi-sweet wines.

"We'll use the Bukettraube for blending into semi-sweet wines," says Charles, "but the intention is to focus on the production of a limited number of quality dry white and red wines. I've planted Pinot noir that will extend our range of cultivar reds. But the Merlot I've planted will be used to establish a quality blended

Fairview

Red wines made in the Fairview cellar made a big reputation in recent years.

red, and I'll attempt to improve that to the position of number one wine on the Estate."

Fairview is a long narrow farm running from high on the decomposed granite slopes of Paarl Mountain to the south-east through a sandy depression—where the vines tend to suffer from lack of water in dry years—to an elevated hill with clay-based soil, which has been found to be some of the best vineyard land on the Estate. This hill, previously planted with Pinotage, has been replanted with Pinot noir and Sauvignon blanc. Charles has found that Pinot noir bears a little too heavily and has had to resort to summer pruning, when the grapes are pea-sized.

"This wine will be lighter than our traditional red style," says Charles, "but so far we've been able to extract plenty of flavour."

Another lighter red from Fairview is Pinotage. Charles has trellised the

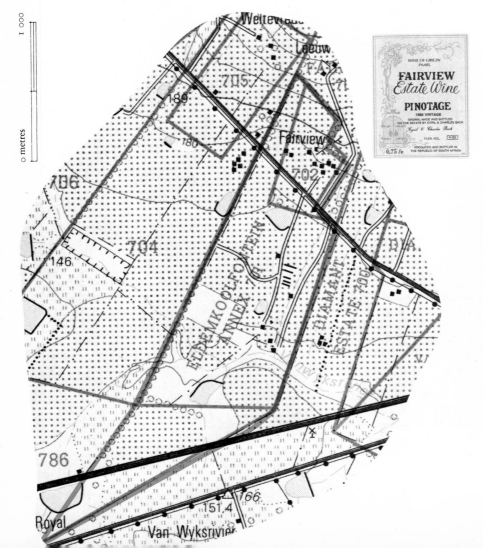

Cyril and Beryl Back have introduced goat's milk cheese, made on Fairview, to the Cape culinary scene.

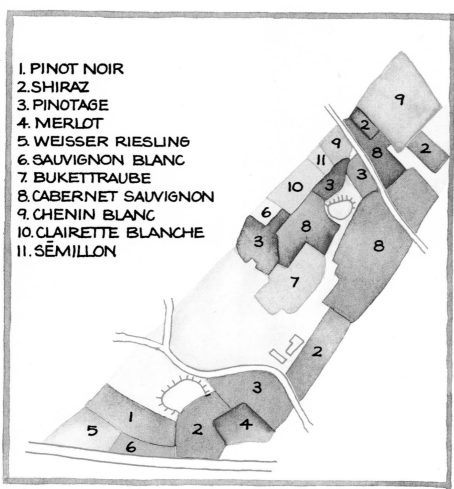

1. PINOT NOIR
2. SHIRAZ
3. PINOTAGE
4. MERLOT
5. WEISSER RIESLING
6. SAUVIGNON BLANC
7. BUKETTRAUBE
8. CABERNET SAUVIGNON
9. CHENIN BLANC
10. CLAIRETTE BLANCHE
11. SÉMILLON

farm's Pinotage and increased the vine's production. In addition to a richly coloured, lighter bodied red wine, he has used the variety to make a very lightly coloured dry rosé.

In general, the higher vineyards on the mountain slopes have Hutton-type, granitic soils and are used to make dry wines. Lower on the farm, the sandy loam soils are excellent sources of sugar and the Steen vines from this area are earmarked for the Estate's Special Late Harvest.

Fairview's other claim to fame is South Africa's only goat's milk cheese. (One third of all the goat's milk cheese produced in France is made on small farms, mostly in or adjacent to wine-producing areas.) Fairview has 300 Saanen milking goats, each producing 75 kg of cheese per year. The goats are grazed on 40 hectares of pasture grown on land unsuitable for vines. The ewes start milking in September and produce an average of 2,5 litres per goat right through to June, when they're given a holiday, before work begins again in September.

The goat's milk is used to produce a compact rounded cheese called Chevin, which is very similar to the French cheese Sainte-Maure, and a Feta cheese. Cyril Back intends to produce a goat's milk yoghurt, in addition to the cheese.

Wine is for sale on the Estate.

Harvesting grapes on the lower section of the farm, in the shallow valley of the Van Wyk's River.

Fairview

Laborie

Though most of the Laborie vineyards are on lower sloping ground, some are grown on the mountain slopes.

(below) Laborie's steeper slopes have been terraced to allow vineyards to be grown under cooler conditions.

When Laborie was developed as a wine Estate, this modern cellar was built directly behind the old cellar.

The advisory and production departments of the KWV, the controlling body for the wine industry, have been harnessed for the development of Laborie as a wine Estate of national and possibly even international status. The KWV cellars produce wine for limited domestic and considerable export consumption, while research and advisory departments assist and encourage grape growers and wine producers to improve their agricultural and wine-making practices. They have a particular interest in the overall improvement of the quality of Cape wines, and are using Laborie to illustrate the benefits of certain techniques and ideas. As the produce of Laborie is bottled and sold, and the Estate is run on business lines, it offers a commercial opportunity to demonstrate these concepts.

Around 1967 the management of the KWV sought premises to house its growing collection of wine industry museum pieces, and land was bought next to La Concorde, KWV's head office, for this purpose. Then in 1972 the historic Laborie property, almost opposite La Concorde on the slopes of Paarl Mountain, was acquired. With its magnificent old buildings, prime position and well-watered, decomposed granite soils, Laborie was ideally suited to the combined concept of a museum and wine Estate.

Laborie was the central member of a row of five long and narrow farms, stretching between the west bank of

the Berg River and the granite outcrops of the Paarl Mountain, all of which were granted to Huguenot refugees who arrived in the Cape during the late seventeenth century.

Isaac Taillefert was a master hatter in Château Thierry, a large town, described then as in Champagne, but today in the district of Brie, home of the famous soft cheese. He married Susanne Briet in 1671 and they had six children. Susanne had inherited vineyards in Monneaux, and from these grapes the family made wine. This was around the period when Dom Perignon and his neighbours were developing the Méthode Champenoise system of a second fermentation in the bottle to produce a sparkling wine. It is not known, however, whether Isaac made still or sparkling wines. In 1687 Isaac, Susanne and the children fled from France to Geneva and travelled to Holland. In 1688 the family left Holland in the company of twenty-four refugees on the *Oosterland* (later to be salvaged by Jan Meerland, first owner of Meerendal, after she ran aground in a storm in Table Bay). After a voyage of two months and ten days, they arrived at the infant Dutch Colony at the Cape.

The Taillefert family chose to settle near other Huguenot refugees on the eastern slopes at the foot of Paarl Mountain, and after receiving official assistance from the authorities in 1690 were granted title to the land they occupied in 1691. The usual size of a land grant at the time was about 55 hectares, and very few settlers were allocated freehold ownership of a single piece of land larger than this. Isaac Taillefert obtained two adjoining farms, providing the family with 110 hectares, by applying for the one in the name of his fifteen-year-old son, Jean. This piece of land was named La Brie, and the adjoining piece, granted in Isaac's name was named Picardie, after the region north-west of Champagne.

The Taillefert family seem to have put the fertile soil of Laborie (which is what La Brie came to be called) and

The mountain side of the Laborie manor house features a traditional Cape Dutch gable and facade.

Picardie to good use. The two farms remained one unit from the time of the original grant until the 20th century. Like all other settlers, they were primarily subsistence farmers and produced most of their own requirements. Surplus products, particularly brandy, were used to barter with those who had taken up specialised professions, such as carpenters.

François Leguat, a French author who visited the Cape and the Taillefert family in 1698, reported that Isaac's home had a beautiful garden in which nothing was lacking, and that the house contained an inner courtyard in which aviaries with a multitude of birds could be seen. He reported warmly on the hospitality of the family and recorded that Isaac Taillefert's wine was the best in the Cape and was as near as possible to the 'inferior Champagne of France'. We can only assume that he meant well.

Isaac Taillefert was granted another farm, in nearby Wellington, in 1699, but it is understood that this was also the year of his death. His eldest daughter, Elisabeth, aged twenty-six, inherited both properties, though her brothers, Jean, twenty-three, Isaac, nineteen, and Pierre sixteen, were still alive. Elisabeth had married Pierre de Villiers, and after her father's death moved with her husband from Groot Drakenstein to take up residence at Laborie.

When Elisabeth Taillefert died, the combined Laborie and Picardie properties were inherited by her eldest son, Jan de Villiers. Jan and his wife Maria, daughter of Johannes Louw, owner of Babylons Toren in Simondium, also bought Neethlingshof, and were visiting a neighbour in the Vlottenburg area when they were murdered by a slave.

Jan's younger brother, Isaac, inherited both the Paarl and the Stellenbosch properties, and after his death in 1767 his wife sold both properties to two Louws from the Drakenstein (Simondium) area, probably brothers of her late sister-in-law. Hendrik Louw became the owner of Laborie in 1774, and when he died a few years later his widow married Johannes Haupt. The original Taillefert family home is believed to have been sited beside the Berg River, and the existing manor house, looking down on Paarl, was built by either Jan de Villiers or Hendrik Louw. The house was extended and improved by Johannes Haupt and most of the adjacent buildings were constructed during Haupt's period of ownership. Laborie was returned to the Louw family in the nineteenth century, and remained in their hands until it was bought by the KWV in 1972.

Determined to preserve as much as possible of this historic property, the new owners found the main house and the Jonkershuis to be in relatively good condition, while the slave quarters and the old cellar were in disrepair. The slave quarters had to be demolished and a guest house, designed to the style of the previous building, was built in its place. The shell of the old wine cellar was retained when the building was restored, and it is now used as a centre for KWV's wine appreciation courses. A new production cellar was constructed behind the original.

With the objective of making classic wine on their small (40 hectares) property, the KWV made a study of the farm's growing conditions to determine the suitability of individual cultivars. It was decided that the Es-

The Laborie complex includes the house (left), guest accommodation (centre) and the restaurant (right).

tate's soils and climatic conditions were admirably suited to the production of light to medium-bodied red wines and 28 hectares of land on the mountainside site were prepared for planting with Cabernet Sauvignon, Shiraz, Pinotage, Tinta Barocca and Merlot. The intention was to produce separate wines from each of these cultivars and to devise from them.

The soils are deep drained and relatively fertile and there is no significant presence of clay. To assist vineyards through drought periods, permanent irrigation has been installed in all blocks. The situation allows a slight cooling effect from the prevailing south-east wind and provides sufficient warmth in summer for adequate ripening. The east-sloping vineyards receive the first rays of sunlight each morning, but fall into shadow early in the afternoon. Most of the Laborie vineyards slope to the north or the south on either side of a central depression running from the mountain toward the river. Other vineyards have been planted on terraces cut into angled, steeply sloping ground, high on the hillside. The broken nature of the ground and the variation in slopes have provided a multitude of different micro-climates and ripening conditions to suit late and early varieties.

As the KWV is the leading supplier of higher-quality planting material for vineyards, Laborie has had the benefit of superb vine stocks. Two different forms of trellising have been used, with the lengthened Perold system used on the terraces on the mountainside and the gable system for the vineyards on more level ground.

When the first small quantities of wine were made, during the late '70s, it was seen that the early sunset on some vineyards and the deep soils affected each variety in different ways, suiting some cultivars better than others, and causing slight changes in the Estate's original plans. Because of the high acidity and slight colour of wines coming from certain vineyards, it was decided to perpetuate Isaac Taillefert's memory with a bottle-fermented sparkling wine, mainly from the juice of Pinotage and Caber-

The well-watered, well-drained soils produce red grapes with high acidity, used to make the base for an Estate sparkling wine.

1. PINOTAGE
2. TINTA BAROCCA
3. SHIRAZ
4. WEISSER RIESLING
5. CABERNET FRANC
6. CABERNET SAUVIGNON
7. MERLOT
8. PONTAC
9. SAUVIGNON BLANC

net Sauvignon grapes. Because this diverted some wines originally planned for use in the red wine blend, a small additional vineyard of Cabernet franc was planted. At the same time, the Estate management decided to add a further product, and now provide a range comprising a blended red, a blended white and a sparkling wine. Vineyards of Weisser Riesling and Sauvignon blanc were planted to provide the source material for the white blend and some of the red vineyards were grafted over to white varieties.

The grapes for the sparkling wine are harvested when not quite ripe (low in sugar and high in acidity), and after crushing all the juice is drawn off the skins as fast as possible to get minimum colour, and fermented at cold temperatures in the Estate cellar. After fermentation is complete, the new wines are given a small dose of SO_2 for preservation and are transferred to the KWV main cellars for the rest of the bottle fermentation process. The base wine is bottled, yeast and sugar added and the bottles are laid to rest for a year, at 10-14 °C. The wines complete the secondary fermentation and obtain extra complexity from contact with the yeast inside the bottle. Then the wine is cleaned, using the transfer or clarification process, as opposed to the *remuage* technique of Méthode Champenoise. The bottles are opened and decanted into a tank under their own pressure, where the contents are mixed, filtered and finally bottled again, retaining most of the original CO_2 dissolved in the wine.

The grapes for the production of red wine are picked mid-ripe (21-23 ° Balling) to make medium-bodied wines. They are crushed and the mixture of juice and skins is fermented at just over 20 ° Balling until the required degree of colour has been obtained, normally between 8 and 12 ° Balling. The liquid must is then fermented dry and the new wines are allowed to undergo a malolactic fermentation. The wines are cleaned and transported to the KWV cellars for maturation in large oak casks, where they remain for more than a year. After ageing, the final blend is made up, bottled and given a period of bottle maturation before sale.

Wine is for sale on the Estate.

Laborie

La Motte

The La Motte property contains a mill built in 1720 and the original Huguenot cemetery.

Looking down the slope of the red vineyards to the cellar, home and the mountains south of Franschhoek.

In 1691 a young German immigrant named Hans Hendrik Hatting arrived in the Cape. He was soon granted burgher rights and was settled in the midst of a group of newly established Huguenot refugees in the Groot Drakenstein Valley. Van der Stel tried to mix Dutch and German pioneer farmers with the French so that the newcomers could prevent the development of a nationalistic enclave and pick up some French agricultural skills at the same time. Hatting was settled among a group of Frenchmen from the La Motte area of Provence, and he named his farm after their home town. His property was next to Pierre (soon to be known as Pieter) Joubert, who was born in La Motte d'Aigues and who named his Drakenstein farm La Provence. In 1695, Hatting lived alone on his 55-hectare property, without a slave or a servant, and his possessions recorded in the census of that year totalled six cattle, a gun and a sword. Pierre Joubert is recorded as married with four children, 5 000 vines, 25 cattle and 200 sheep.

In 1705 Hatting moved to the farm Goede Hoop, and in 1709 he sold La Motte to Joubert. In 1712 Hatting bought a farm in Stellenbosch from Arnout Jansz and may have been responsible for renaming it Spier (Hatting was born in Speyer), and lived there until his death in 1728. The family appear to have fallen on hard times as his son asked for a freehold grant of a farm in the Paardenberg area, for no payment, in 1748. He claimed to be supporting his poverty-stricken mother and her children.

Pieter Joubert, the second owner of La Motte, was no stranger to poverty, having grown up in rural Provence during an extended period of material hardship, as well as religious intolerance. The more zealous Protestants fled their homes throughout France, but the large proportion of refugees from Provence were stimulated to move by the depressed economic condition of their rural area. An English visitor to Provence in 1676 reported the mostly illiterate families to be reduced to eating the bark off trees. Pierre Joubert, Pierre Mallan, Susanne Reyne and Isabeau Richard left Provence in 1686, making their way to Protestant Holland via Geneva. Mallan married Richard and Joubert married Reyne shortly before the *Berg China* left Europe for the Cape—a perilous voyage that lasted several months. We must presume that Pierre Mallan and Susanne Reyne perished during the voyage, as Isabeau Richard was described as the wife of Pierre Joubert when the ship docked at the Cape. Fifty of the *Berg China*'s passengers required hospitalisation on arrival.

Though Pierre and Isabeau lacked formal education, the literacy of other refugees encouraged them to provide education for their children, and the soldier Jean de Camau of Toulouse was assigned to the Joubert household as schoolmaster in 1706.

Like all the other pioneer settlers, Joubert was given assistance by the Company until he was able to support himself, and early in his Cape career he is recorded as owing 230 guilders, and goods worth 405 guilders. His rural background proved to be of assistance, and he became an important farmer, with five properties in Drakenstein and Wolseley. After buying La Motte from Hatting, he acquired the adjoining farms Bellin-

The original La Motte home was destroyed in the nineteenth century and a new home was built in 1836.

champ (later to be known as Bellingham) and L'Ormarins a few kilometres away, and obtained a share of the lease allowing retail sale of wine in Cape Town. In 1714 a neighbouring farmer Gilles Sollier, who spoke English and Dutch in addition to French, arranged the purchase by Pieter Joubert of two 'Madagascar negro ladds' imported to the Cape on the English vessel *Delicia*.

In 1715 a marauding band of 'Hottentots' stampeded Joubert's stock on his Wolseley loan farm and made off with half his cattle. He petitioned the administration for the ownership of the property in compensation for his loss, and it was granted to him in the following year. By 1730, shortly before he died, he had 19 slaves and farmed with 1 000 sheep, 100 cattle and 20 000 vines on his various properties. His cellar contained 26 leaguers of wine.

Many farm mills were built in the pioneering days for the conversion of grain into flour. Joubert is believed to have constructed a mill house on La Motte in 1721. Unfortunately, only the hursting and three of the mill stones remain. In 1974 Dr. Anton Rupert, the present owner, brought mill machinery from an old Ceres mill and re-equipped the La Motte mill and mill house.

La Motte also contains the old Huguenot cemetery, where Joubert and many of his fellow refugees were buried.

Like many Cape properties, the La Motte farm was not used to full potential over the next two and a half centuries. In 1825 Gideon Joubert had 39 000 vines on the farm and 14 leaguers of wine in the cellar. On La Provence, next door, Hermanus de Villiers had 160 000 vines and 40 leaguers of wine.

When Dr. Anton Rupert attended the 1970 auction sale of the property (he had bought L'Ormarins, a few kilometres away, the previous year) he had not intended to buy La Motte, but because the highest bid was less than his idea of the true value, he made the successful bid for the farm. Over the next few years, he spent almost as much as the purchase price on the restoration of the buildings.

Anton Rupert was born in Graaff-Reinet, the son of a lawyer. From an early age he took an interest in manufacturing, and eventually became one of the world's leading producers of cigarettes. Based in the western Cape, he soon expanded his interests into liquor, and in 1962 introduced the first two private-cellar wines to carry the concept of Estate wines on the label, Alto and Theuniskraal, on to the South African market. With the purchase of his two Drakenstein Valley farms in 1969-70 he began to lay the foundations for his own family Estates. Anthony, his son, studied at Geisenheim, worked on wine Estates in Germany in the mid '70s, and returned to South Africa to develop the two Estates in 1978, and the preparation of land for the planting of classic cultivars was begun.

La Motte stretches from the low foothills of the mountains down to the Berg River, not far from Franschhoek village. The vineyards of Sauvignon blanc, Chardonnay and Weisser Riesling are planted on level alluvial soils beside the river, and the blocks of Cabernet Sauvignon, Merlot, Cabernet franc, Pinot noir and Carignan are planted on the south-facing slopes at the base of the mountains.

The average annual rainfall in this central part of the Franschhoek valley is about 900 mm, but all the vineyards receive supplementary water in spring and summer. Because La Motte has moderate water resources, only certain vineyards are provided with permanent irrigation, with the balance receiving sprinkler irrigation during summer. The soils on La Motte lack clay and lose moisture during the late summer, making ripening an arduous process for the late varieties. More recent planting has concentrated on earlier-ripening varieties, with an emphasis on white wine varieties. Like most coastal Cape vineyards, La Motte's soils lack lime, and Tony Rupert has added large quantities of lime to soils being prepared for the planting of vines. The climate is warm, with grapes ripening three to four

The Estate's red variety vineyards are on the lower foothills.

days ahead of grapes belonging to the same variety on L'Ormarins in the Groot Drakenstein area, only five kilometres away to the west. Afternoon and evening temperatures during summer are moderated by the influence of a westerly wind that blows through the Helshoogte gap between Simonsberg and Jonkershoek and courses up the Franschhoek Valley past La Motte.

The deep-drained soils influence the style of wine made from La Motte's vineyards. The grapes are picked relatively early, to make light, fruity wines for early consumption. Because Tony Rupert's ambition is to make classic wines, the vines are winter pruned to severely restrict their crop, and excess bunches on both young and producing vineyards are cut off in summer. This practice helps them to produce optimum quality wines.

Wine is not for sale on the Estate.

Landskroon

The De Villierses of Landskroon are like the Paarl rocks behind their farm, as they too have been there for a long time. Indeed, the Paarl Mountain has a very strong influence on their lives, with the Landskroon wine Estate situated on its south side facing west. The sun rises behind the mountain and the farm gets little early morning sun. This helps the grapes on the vines planted high on the hillside to ripen slowly, and is likely to be one of the reasons for the excellent quality of the wine made on the farm. The mountain is a continuation of a long granite chain that runs north from Somerset West to Stellenbosch and then on to Paarl. Most of Landskroon's vineyards are situated on the lower slopes of Paarl Mountain and, in this type of soil, the vines that do best are the red varieties, so Landskroon is primarily a red wine Estate.

The De Villierses have been in South Africa for a long time. Three brothers emigrated with a shipload of French Huguenot immigrants in 1689 and were given a tract of land to farm at Franschhoek and plant vineyards. The youngest brother, Jacques, later acquired the farm Boschendal, in the Groot-Drakenstein district. He planted vines near the river, irrigated them and established a cellar. The wine he made was transported from Groot-Drakenstein to Cape Town for sale, which entailed a two-day journey by ox-wagon.

The Landskroon farm also has a long tradition of wine making. One of

Paul de Villiers has been the Estate vintner since 1962. His son, Paul, is now the Landskroon cellarmaster.

Looking across the Van Wyk's River and the Paarl Valley toward Simonsberg.

Landskroon 75

the first farmers to settle on the south side of Paarl Mountain called his farm Landskroon and in 1692 planted a small number of vines of unknown origin and type.

In 1872 Paul, a Boschendal De Villiers, bought a portion of the original Landskroon farm, named Weltevreden. He was the great-grandfather of the present vintner (also named Paul), and was the fifth generation De Villiers to live and farm in South Africa. He named his first-born son Paul, to whom we must refer as Paul the second, because he too named his first son Paul. Tradition declared that Paul the third should also name his first-born Paul, and this is the man who now runs the Landskroon wine Estate. Between 1872, when the first Paul de Villiers bought Weltevreden, and 1963, when Paul the fourth took over, the family purchased two adjoining farms, one with the original name of Landskroon and the other named Schoongezicht. The three properties have been amalgamated and are now farmed as one unit under the name Landskroon Estate.

When Paul the first came to the farm, he found a substantial area planted with vines, but their type is unknown. He then planted Cinsaut, Steen and Muscadel vines to make sweet fortified wines, and built a cellar on the farm for that purpose. Like his great-great-grandfather Jacques, Paul the first carted wine from the farm to Cape Town for sale. It brought good prices from wine merchants for both local consumption and export.

Paul the second was in charge of the farm when the KWV was formed in 1919. He quickly saw the advantage of selling all his wine at regular prices to the KWV, instead of relying on the fickleness of the wine-consuming public. Until 1973, almost all the farm's wine was sold to the KWV each year. Paul the fourth, the present owner, has been the first to break this custom, by retaining some of his better-quality wine for sale under his own Landskroon label.

Paul the third was probably the most famous of all the wine-farming De Villierses. During his time in charge of the farm he formed an association with Dr. Niehaus of the KWV in the development of sherry-type wines. Dr. Niehaus was busy trying to make sherry-type fortified wines that could compete on a quality level with the fortified wines of Spain. While Dr. Niehaus was trying to find a natural South African yeast that would make a 'fino' sherry-type wine, he used part of Paul the third's cellar as a laboratory. He was also busy trying to find the elusive breed of yeast on other farms in the Stellenbosch–Paarl basin, but the Landskroon cellars contained for many years some vats of sherry-type wine at different stages of fermentation. Eventually, as is well known,

The De Villiers family moved to Landskroon over 100 years ago and now the fifth-generation Paul de Villiers is making wine in the Estate cellar.

Though Pinot noir produces wine with a lighter colour, Landskroon's wines are generally dark and full-bodied.

A reliable producer of good-quality medium to full-bodied wine, Pinotage is highly valued on Landskroon.

On Landskroon's moderately fertile soils Pinot noir grows vigorously and produces a crop of healthy grapes.

Grown in the sandier soils in the base of the valley, vines produce smaller crops of excellent quality.

Facing south-west, Landskroon enjoys long periods of sunshine each day and receives the cool south-west wind.

Dr. Niehaus found the 'flor' yeast on Landskroon and other farms.

Later the KWV encouraged the third Paul de Villiers to plant the Portuguese varieties that had proved to be best suited to South African conditions for the making of ports. Amongst these were Tinta Barocca, Souzão and Tinta Roriz, which are still in production on Landskroon.

Like his father and grandfather before him, Paul the third produced some dry table wine, most of which was red. This was unusual, for almost all of South Africa's wine was white and had always been so, but Paul the third continued to develop the production of the Portuguese red-skinned varieties, and also began to plant Cinsaut in quantity for the production of port wines. He found Cinsaut to be unexpectedly successful in the production of medium-sweet red fortified wine on this particular farm.

Planted on the slopes of the mountain, with little afternoon sun, and regularly getting only two-thirds as much rainfall as vines grown in Stellenbosch, Cinsaut developed a hardy, small, dark-skinned berry on Landskroon. It produced less juice but during fermentation the must quickly took rich colour and flavour from the skins and undoubtedly helped Paul the third to become one of the most famous port-type wine farmers in the country. Because of the involvement with production of medium-sweet port-type wines, Landskroon has had more red vines that white for as long as the present owner, Paul the fourth, can remember.

"We've always had a ratio of about 75 to 25 per cent, red to white. The granitic-type soil and our position on the mountain seem to be better suited to red wine varieties than to white. I worked with my father from 1962, and in that time we planted more and more Cinsaut. We also kept an area of vines under Steen, White French and Green Grape for the making of sherry and white ports, to help keep our harvesting season even, but we must have had seven or eight red varieties to our three white varieties during those years."

Landskroon

Landskroon is a very long farm. This is a view from the mid section, looking north toward the cellar and home.

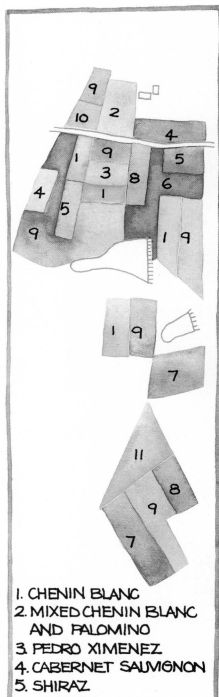

Paul the third left the farm to his two sons, Paul the fourth and Hugo. Like many other South African wine farms Landskroon also contains a dairy, and the property is run on a partnership basis with Hugo taking care of the dairy and Paul totally involved in wine making.

When Paul the fourth took over the care of the vineyards in 1962, his production was almost entirely port- and sherry-type wines. Like his father before him, and all the De Villierses before that, he did not sell grapes or grape juice but made his own wine. Since Paul has been running the farm, he has planted Pinotage, Shiraz, Cabernet Sauvignon, Cabernet franc and Pinot noir. Since 1969 replanting has been aimed at improving the quality of dry table wine production.

The farm has become one of South Africa's premier producers of Estate red wines. The decomposed granite soils of Landskroon's south-facing vineyards on Paarl Mountain, and the sandy soils that are found lower down the farm assist red varieties to reach high sugar levels and make full-flavoured, heavy-bodied red wines. Paul de Villiers was one of the first of the adventurous quality wine producers, planting uncommon but classical red varieties before his competitors. To add variety to the standard range of Cabernet Sauvignon, Pinotage, Cinsaut and Tinta Barocca, he planted Pinot noir and Cabernet franc. In 1978 he bottled Pinot Noir for the first time and followed this with his maiden vintage of Cabernet Franc in 1979. Both wines were dramatically successful and the size of these vineyards has been increased to cope with the demand. In 1981 he won the prize for the champion red wine of the show at Paarl with a 50/50 blend of Cabernet Sauvignon and Cabernet Franc. He narrowly missed repeating the performance in 1982, when his blend was among the finalists.

Paul sells six red cultivar wines and one blend, Bouquet Rouge, a blend of his best Cabernet Sauvignon, Shiraz and Cinsaut. "I don't understand what South African wine drinkers are looking for," he says. "This wine is made from my best reds, and yet customers prefer to take the pure cultivar wines that are probably less complex. They seem to believe that blends are necessarily inferior. My greatest red wines will always be blends, I feel, but until my customers recognise the quality I'll only be able to offer limited quantities of these wines."

In 1980 Paul's son, also Paul, came into the cellar and now operates as cellar master. Paul the fifth brought an interest in dry white wines and the cellar has been modified to produce fragrant, full-bodied dry wines, from Steen in particular. Sauvignon blanc has been planted to allow the younger Paul to make a range of quality dry whites matching the Estate reds.

Wine is for sale on the Estate.

L'Ormarins

Commenced by a Huguenot, completed by Johannes Marais in 1822, the home was bought by Dr. Anton Rupert in 1969.

Between 1688 and 1700 approximately two hundred French Huguenot refugees arrived at the Cape, survivors of religious persecution in France and the six-month voyage from Holland. Among the first group, in 1688, were a number of farmers originating from the southern region of Provence, where fertile soils have provided France with fruit and grapes for many centuries. The Dutch East India Company allowed each to select a small portion of land in the Stellenbosch-Paarl area, and lent them spades and other farming implements, including a part share of a plough and tools to build and maintain a hut. The company also lent each pioneer some cattle to assist his enterprise. The loans were to be repaid at the first opportunity. Some of these hardy Frenchmen were married, like Pieter Joubert, who married twice between Holland and Groot Drakenstein (it is presumed that his first wife died). Others like Jean Roi and Paul Roux were bachelors and had difficulty in persuading women to share their lives in the bush below the towering mass of Groot Drakenstein, where wild animals caused the farmers to select pieces of land within sight of established habitation. The valley where they settled was known as Olifantshoek before they arrived. Naturally it soon became known as Franschhoek. The district tax imposed from the castle at Table Bay was known as 'lion and tiger money', as it paid, among other things, for a protection force located in the valley with the farmers.

Jean Roi and Paul Joubert were given adjoining pieces of land on the slopes of Groot Drakenstein. Roi named his farm L'Ormarins, while Joubert chose the name La Provence. Because Roi's friend, Paul Roux, is known to have come from the village of Lourmarin, one can suppose that this also applied to Jean Roi. In the census of 1692, Roi was declared to be living alone on his tiny farm. He had planted 4 000 vines and had reaped 30 bags of wheat from the three bags he had used as seed. He also had four oxen, one cow and one heifer.

In 1694 Roi's industrious farming prompted Simon van der Stel to give him freehold ownership of L'Ormar-

The Great Dragon Mountain (Groot Drakenstein) can be seen behind the now terraced slopes.

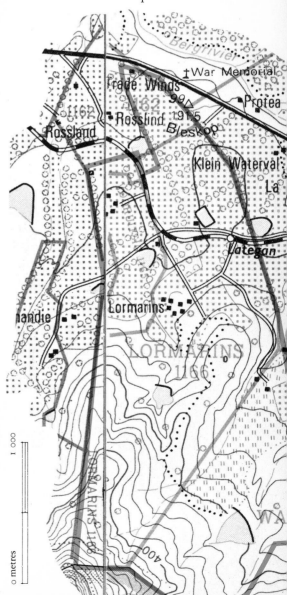

ins. Twenty years later it was discovered that the official grant had not been made as promised and the oversight was rectified.

The Huguenot citizens of the southern end of the Paarl Valley were able to persuade Simon van der Stel to bring out a French protestant minister to provide spiritual guidance for the refugees. They built a church on ground beside the farm Babylons Toren, and a curious tale involving the church, a storm and Jean Roi has been passed down. During a particularly savage north-east storm the church building collapsed and Jean Roi found himself in possession of the congregation's imported church bell. The bell was found in the fork of an old oak tree at L'Ormarins by one of the farm's twentieth-century owners, and is now mounted between concrete pillars, in the form of an old Cape slave bell.

The loneliness of the hillside pioneer's life came to an end in 1712, when Jean Roi married Marie le Febre, widow of Gabriel le Roux. Roi died in 1720 and Marie and her nine children ran the farm until she married Frans Haarhof, a Dutch East India Company soldier, whom she had employed as a part time labourer, in 1724. After three years Haarhof sold the farm to Pieter Joubert, his neighbour on Bellinchamp (later to be known as Bellingham).

It is believed that at that time Joubert was one of four lessees of the right to sell wine and brandy in the Colony at Table Bay. In 1727 a townsman, Hendrik van Dijk, requested permission to take over Joubert's share of the lease and the premises. Joubert was a successful farmer and businessman, leaving six farms, including L'Ormarins, Bellinchamp, La Provence and La Motte, to his wife on his death in 1732. It is not known on which of his farms Joubert lived during his later years, but it is believed that the oldest house on L'Ormarins today was built during his period of ownership.

The period of greatest prosperity in the early Cape began toward the end of the eighteenth century. The old cellar at L'Ormarins bears the date 1799 and the main house carries the year 1811. These buildings were either constructed or renovated by Johannes Marais, whose father had married the widow of Johannes de Villiers, son of the owner of Boschendal. The De Villiers family had owned the farm for 32 years. Marais's son Isaac was one the Cape's leading wine makers and distillers. He farmed L'Ormarins for 38 years and from its vineyards made Madeira wine that won the Cape Championship in 1833, and was runner up ten years later. In 1845 his Pontac wine won the £20 trophy, and in 1848 L'Ormarins brandy won the Colony's trophy. It is believed that Marais lived alone on the farm and left no will and no heirs, and so L'Ormarins was sold 'out of the family'.

The farm was allowed to run down over the following decades. Both the next two owners were declared insolvent, and a succession of wealthy families bought the property, chiefly because of the magnificent house, and L'Ormarins became more properly a residence than a farm. Between 1919 and 1969 the property was owned by, among others, the Duke of Abercorn, Countess Cienska and Lady Vincent. Wine production in the old cellar ceased in the early '60s.

In 1969 the property was purchased by the Rupert family with the intention of developing it as a wine Estate. At that time the run-down vineyards contained Cape Riesling and Steen vines and the major income came from peach orchards. During the grape harvest of 1971 the crop of the Cape Riesling vines was picked and transported to a neighbour's cellar where the grapes were crushed and the juice fermented. The wine was de-

Though L'Ormarins has vineyards of four noble red varieties, the Estate concentrates on top whites.

The hillsides of L'Ormarins have been terraced to allow vineyards to be cultivated and provided with irrigation.

livered to the Bergkelder, where some of it was used to make a special batch of sparkling wine and the rest bottled as still table wine. These wines caused a lot of interest and fuelled the desires of Dr. Anton Rupert and his son Anthony to re-establish L'Ormarins as a force in the Cape wine industry. Anthony studied viticulture in Europe before returning to live on the farm and begin redevelopment in 1978.

The first stage involved the clearing of 100 hectares of bush on the higher slopes of the farm. This steep land, below the granite and sandstone cliffs of Groot Drakenstein, was terraced to allow the planting and cultivation of vineyards. The light, loamy soil was found to be deficient in several macro-elements and required a major addition of lime and phosphate. Bulldozers were hired to plough the corrective fertiliser deep into the ground and to mix the reformulated soil thoroughly.

Lying in the shadow of the tallest mountain in the region, L'Ormarins receives 1 500 mm of rainfall each year, mostly during the winter, but the steeply sloping land was found to be lacking in moisture by late summer, and Anthony decided to build a major dam in the catchment valley below the famous Groot Drakenstein waterfall. This dam and several smaller ones now contain almost 850 000 000 litres of water and supply the micro-spray irrigation system that has been installed in every vineyard on the farm.

The old Riesling and Steen vineyards have been removed and the whole farm has been provided with new or replacement vineyards totalling about 200 hectares, since 1978. More than 80 per cent of the vineyards have been planted to white varieties. "This farm is best suited to early ripening varieties," says Anthony. "We face eastward and get the early morning sun over the Franschhoek Mountains and the whole farm drops into shadow an hour or more before the west and south-west facing farms of Paarl and Stellenbosch. Most of the early-ripening classic and high-quality cultivars produce white grapes, and so we've planted many more white vineyards."

Anthony has planted, or is planting, in descending order of quantity, Sauvignon blanc, Chardonnay, Rhine Riesling, Cape Riesling, Chenin blanc, Pinot gris, Sémillon and Gewürztraminer vines. The red varieties chosen are Cabernet Sauvignon, Merlot, Pinot noir and Shiraz.

"We're very high. Our temperatures, particularly at night, are relatively cool and we have fewer hours of sunlight than most Estates. We have to choose varieties that can give flavour and delicacy in their wines. We'll have to be very careful with our reds. We can only hope that we've chosen the right slopes and the right cultivars. Except for Pinot noir, our other reds are middle to late season varieties. We've planted these on our

A Chenin blanc vineyard in a sheltered basin behind the house.

Cultivating a young vineyard of Rhine Riesling on the hillside beside the cellar.

The modern pressing cellar of L'Ormarins was built on the slope.

L'Ormarins

warmest soils, on the lower slopes of the farm, and the next few years will at least give us an idea if we've guessed correctly."

Because of the hilly nature of the mountainside farm, L'Ormarins provides a selection of altitudes and a choice of slopes that face each of the points of the compass. In some instances, where the terraces run around the circumference of a hill, each row of a vineyard faces several different directions. One Pinot noir vineyard has 480 metre rows. Sections that share similar growing conditions will be harvested and fermented together and will provide different tanks of wine that can be compared, looking for the ideal.

The Estate is dedicated to the pursuit of quality and the production of outstanding South African wines. Vines are planted closer together than has been the custom in South Africa, to provide more vines in a given area, and to expect less yield from each vine. Accordingly, all vines are severely pruned to allow each to concentrate flavour substances in a limited quantity of grapes. Though all vineyards are provided with irrigation, the policy precludes the addition of water during the ripening of the grape.

Anthony's first real season was in 1982 when his three and four-year-old vineyards provided his new cellar with 600 tonnes of grapes. The cellar had not been completed in time for the 1981 season and the first harvest from the three-year-old vines were

The summer south-east wind condenses into cloud over Franschhoek Mountain but is little felt on sheltered L'Ormarins.

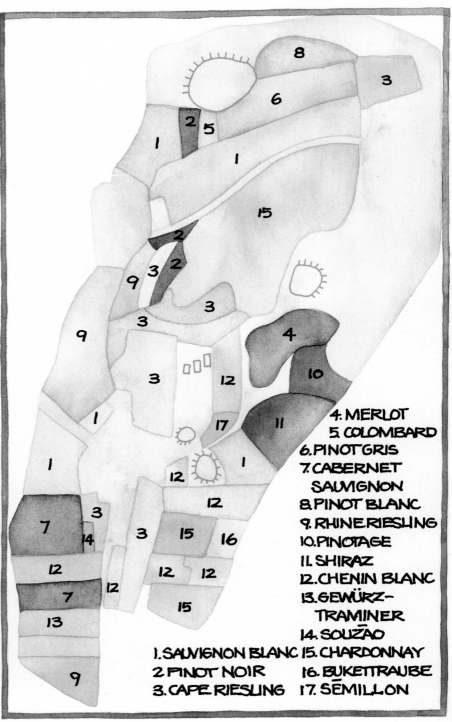

1. SAUVIGNON BLANC
2. PINOT NOIR
3. CAPE RIESLING
4. MERLOT
5. COLOMBARD
6. PINOT GRIS
7. CABERNET SAUVIGNON
8. PINOT BLANC
9. RHINE RIESLING
10. PINOTAGE
11. SHIRAZ
12. CHENIN BLANC
13. GEWÜRZTRAMINER
14. SOUZÃO
15. CHARDONNAY
16. BUKETTRAUBE
17. SÉMILLON

delivered to a wine merchant. In 1982, a warm year, L'Ormarins grapes reached full ripeness with plenty of acidity and made some very interesting maiden-crop wines. The Rhine Riesling was judged class winner at the regional young-wine show and the Sauvignon Blanc won a gold medal in its class. The Sauvignon Blanc was aged in new barrels of Nevers and Limousin oak before release.

Anthony believes that the American-developed clone of Sauvignon blanc, with its flowery aroma and high proportion of acidity in the juice, should be grown alongside an alternative French clone with more distinctive flavour. The two wines could be blended to take advantage of the benefits of each member of the family.

After crushing, most of the white grapes are given a period of skin contact, where the broken skins are left to soak in the cold, unfermented juice. Sauvignon blanc and Chenin blanc, for instance, receive skin contact, while Cape Riesling skins are immediately separated from the juice. The juice of the white grapes is cold settled to remove impurities and then inoculated with a culture of yeast to begin fermentation. The white musts are fermented at between 12 and 14 °C. The red musts are given an extended fermentation on the husks at controlled temperatures, to enhance the elegance and delicacy expected from grapes grown under cooler conditions.

It is intended to mature all the wines to be sold under the L'Ormarins label in small oak barrels in the Estate cellar. The beginnings are very promising and it is hoped that L'Ormarins will continue to enhance the name of the Groot Drakenstein Valley.

Wine is not for sale on the Estate, as national distribution is handled by a Stellenbosch wholesaler.

The original vineyards were planted on the lower slopes. Tony Rupert has terraced the high slopes for new plantings.

Most vineyards on L'Ormarins slope north toward the sun, but benefit from breezes on the high slopes.

Villiera

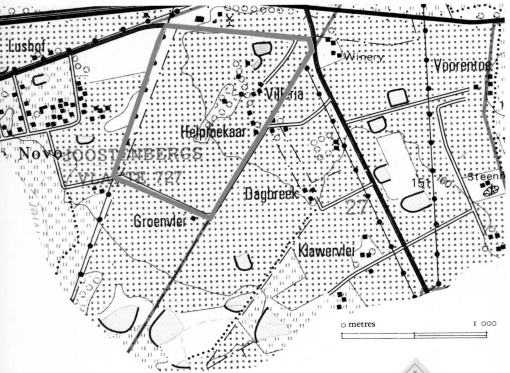

Villiera is on the Koelenhof border between Stellenbosch and Paarl.

Once a notable producer of white wine, sought after by wine merchants for fine quality Steen, Villiera went through a period of depression during the '60s and early '70s when the vineyards declined and the cellar was allowed to fall into disrepair. But in 1975 the farm was bought by Helmut Ratz, a visiting Austrian, and Villiera began to climb back towards the forefront of Cape farms.

Villiera stands on the north-western edge of the Stellenbosch–Bottelary area, on the natural boundary of the land that is used commercially for vine growing. During the nineteenth century a large tract of the best land to the west and north of Stellenbosch was owned by the famous Beyers family. They owned Kleigat (Knorhoek), Driesprongh (Delheim), Muratie, Uitkyk, Muldersvlei, Elsenburg and others. These farms bordered an expanse of sandy grazing land to the west, known as Joostenbergvlakte. In 1893, some 575 hectares of Joostenbergvlakte were sold by the Colonial Government to the inheritors of the Estate of Jan Andries Beyers for £209.

Approximately 120 hectares of this land was sold in the 1930s to J. W. S. de Villiers, who named the new farm Villiera. De Villiers planted vines in the sandy soil and sold the grapes he grew to merchants in Stellenbosch. In 1941 he built a cellar on Villiera and began to make wines from Cinsaut and Steen. The vines, as was the custom, were untrellised and watered only by nature. De Villiers built up the production of the farm by fertilisation and skilful pruning. In 1950, he sold Villiera to C. P. Naudé, who harvested more than 1 000 tonnes from Villiera in his first two seasons. But this was also the beginning of Villiera's decline. Though Stellenbosch wine merchants remained prepared to pay good prices, production fell below 300 tonnes, and wine was no longer made in the cellar. At this point Helmut Ratz bought the farm.

Ratz was an entrepreneur from Kärnten near the Yugoslavian and Italian borders, in Southern Austria. He first came to South Africa on a hunting trip in 1972. Later, he returned with plans to invest in the Stellenbosch area. He wanted a farm that could support itself on the South African market, and at the same time supply his Austrian hotel with a quality wine from a far-off, exotic country. In 1975 he brought Josef Krammer, an Austrian wine maker with several years' experience in South African cellars, to help choose the most suitable farm. Together they chose Villiera, and Josef was installed as the manager and wine maker.

In 1975 most of Villiera's vineyards contained Cinsaut, with the balance mainly Steen. A small vineyard of

Preventative spraying and dusting programmes during spring and summer protect the health of the vines and grapes.

An altitude of only 15 metres separates Villiera's highest and lowest vineyards.

young Cabernet Sauvignon vines had been planted in a clay hardpan patch of ground, and was struggling to survive. Josef set out to make his first season's wines from the unpromising vineyards. Most of his 1976 crop was sold to a merchant. A small quantity, about 7 000 bottles, was sent to Austria to test the reaction.

He instituted a programme of vineyard redevelopment, financed mainly by wine sales. In the first seven years of Austrian ownership, much of the Cinsaut was removed, and a soil reconstitution programme was commenced. About 80 per cent of Villiera's vineyards have sandy soil, with little evidence of loam. The balance have a predominance of clay. Both soil types present problems. The sandy soils are easily leached and become deficient in trace elements. The clay-based soils present a hard-packed barrier to the root development of the vines, and have to be ripped deep to allow healthy growth. All of Villiera's soils are acid and have to have had seasonal fertilisation to correct deficiencies.

He has replanted more than 20 per cent of the farm with quality new varieties and the programme is continuing. "The white varieties seem to do very much better on this farm. It's quite warm here and early-ripening varieties seem to have an advantage. I like to make light, easily drinkable wines that other people can enjoy. Steen, Cape Riesling, Rhine Riesling and Sauvignon blanc ripen easily, with plenty of acidity. But I like to pick the grapes early so that the wines are light, fresh and have plenty of fragrance. Chenin blanc does very well here, and is a very good blending partner for the two Rieslings."

Josef has attempted to solve the farm's water problems in drier years by the construction of several shallow dams. The altitude difference between the highest and lowest points on the farm is only fifteen metres and the dams have trouble in retaining enough water from the limited catchment. The water is only used to assist the growth of young vines during very hot periods. The rest of the vineyards on the farm are not irrigated.

Josef was a pioneer of light-bodied red wines in the Stellenbosch area. His Shiraz-based Alfresco and Cabernet-based Rubiner blends are gentle, soft wines and are suitable for summer drinking. His whites—the Steen-based Operette blend, Rhine Riesling and Sauvignon Blanc—have rather more fruit and depth of flavour. The farm is likely to build a greater reputation for whites than for reds.

Each year since 1976 Josef has sent Helmut Ratz in Austria between 10 000 and 20 000 bottles of the Estate's best wines, white and red, for sale in his hotel. The wines have proved popular, selling for three times their South African prices.

In 1979 Josef started offering visitors during summer months a cold lunch featuring Austrian specialities. The Villiera alfresco lunch contains meats smoked on the farm and salads and vegetables grown on the property.

Wine is for sale on the Estate.

Welgemeend

The couple climbed out of the car and walked slowly across the pebbled ground. They looked around the tiny farm wedged between the freeway and the old highway to the north. There were a few thousand vines, unkempt and untended, in ragged rows with roughly in the centre of the property a small farmhouse and a few outbuildings. Their plans were small, their means were modest and it looked like a pretty good place to start. All they wanted was a place in the country to live, somewhere to grow a few vines and make a little red wine.

Billy Hofmeyr, a land surveyor by profession, won a national wine tasting competition in 1974. Shortly thereafter he and his wife Ursula put their savings together and searched for a smallholding with a wine quota, where he could indulge his dreams and she could extend her garden. The 14-hectare piece of land they found stands on a ridge adjoining the southern extremity of Landskroon, one of Paarl's top red wine Estates. During his first winter, Billy planted small blocks of Cabernet Sauvignon and Pinotage vines, separated by the farm road that runs down the centre of the property. He then obtained a few hundred vines of Merlot, Cabernet franc and Petit Verdot, and made a small mother plantation from which he selected cuttings to establish vineyards beside the Cabernet Sauvignon in later years. He subsequently planted blocks of Shiraz and Grenache beside the Pinotage, and the recipe for two different styles of blended wines began to develop.

Following the classic recipe for a top-class blended red, Billy has planted the major part of the property with 70 per cent Cabernet Sauvignon, 20 per cent Merlot and Cabernet franc, and 10 per cent Malbec and Petit Verdot. His second vineyard, when completed, will have equal parts of Pinotage, Shiraz and Grenache. Conditions of life for these vines are rather harsh. There is a little sand in the surface layer of soil, with some humus. The soil mixture has a gravel base with pockets of sand, and stands on a platform of clay that varies greatly in depth, showing on the surface in some parts, but is mostly between one and two metres below the surface. Where the clay approaches the surface, Billy has had to break it up to get the vines to grow. Two small dams provide water for the establishment of young vines in this rather severe environment.

Grapes are picked in lug boxes to prevent premature crushing.

Rainfall on the Welgemeend Estate is inconsistent, but the average of 600 mm per year is sufficient for the vines to grow and produce 6 to 7 tonnes of grapes per hectare per year when mature. The clay holds sufficient moisture in the autumn for the vines to ripen the grapes without significant loss of acidity.

Billy has found that he has chosen, quite by accident, a farm with a number of natural advantages for the growing and ripening of Cabernet Sauvignon. The texture and geological structure of the soil are similar to those of St. Émilion in Bordeaux. The Bordelais prefer not to irrigate their vines and Billy has insufficient water to do so. Cabernet Sauvignon in Bordeaux produces 6 to 7 tonnes of grapes per hectare, just as it does at Welgemeend. The farm is situated on the crest of a ridge, in line with the south-east wind coming through a gap between the Simonsberg's foothills and Klapmutskop. This prevailing summer afternoon wind moderates the temperature and gives the vines cool nights through most of the ripening period.

Harvest time coincides with that of Paarl and precedes the majority of Stellenbosch Estates by between one and two weeks. The summer rains that cause hazards for Stellenbosch's

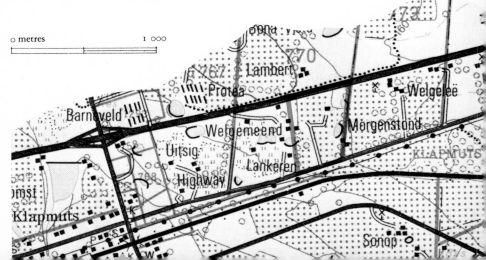

86   Paarl

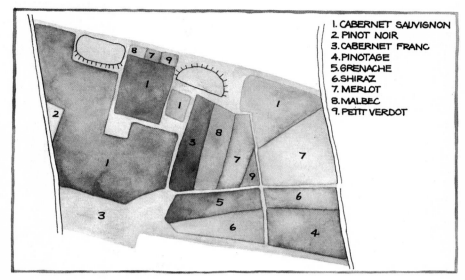

Once sufficient colour has been extracted, skins are pressed in the hydraulic basket press to obtain every drop of flavour.

Billy Hofmeyr, the Cape's first producer of a Bordeaux-blend red.

Fourteen hectares of red varieties take two weeks to harvest.

later-ripening Cabernet Sauvignon have not yet been a problem for Welgemeend. Billy's Cabernet Sauvignon does not all ripen at the same time, and because the farm is small he and Ursula are able to bring in all the crop at close to Billy's chosen degree of ripeness over a period of about ten days. The moment for harvesting comes when the grapes have an average sugar content of 21,5 ° Balling, regardless of pH and acidity content. The pH of the grapes is always low at this point and the low pH factor is retained by all the wines.

The Merlot, Cabernet franc, Malbec and Petit Verdot are picked and crushed together. The Cabernet Sauvignon reaches the same degree of ripeness about ten days later and is fermented separately. The mash of skins and juice is fermented with a culture of dried yeast, at a temperature of about 25 °C. Billy's first wines, in 1979, were fermented cooler, but turned out to be very delicate. He believes he extracts greater flavour without sacrifice at the higher temperature. The sugar is allowed to reduce to about 3 ° Balling before the skins are removed and pressed in an hydraulic basket press. All the pressed liquid is returned to the fermenting must and the combination is allowed to ferment dry in a closed tank. After fermentation has ceased, the wine is allowed to settle. When clean, it is put into small oak casks to mature and gain some extra character. One third of the Welgemeend wood is replaced each year with new oak from France, so each Welgemeend Cabernet blend benefits from contact with new wood. This wine remains in wood for at least a year, after which it is bottled and released to the market under the name Welgemeend.

The Estate's second wine, Amadé, is made from the vineyards of Grenache, Shiraz and Pinotage. The grapes are picked and crushed together. The fermentation technique is similar to that used for the Welgemeend wine but the wood maturation period is appreciably shorter. When casks have been used for three years to mature the Cabernet blend, Billy has them scraped out by a cooper and uses them for one further season, in which Amadé is given 3 to 4 months' wood ageing, after which they are sold.

The size of the farm and Billy's surveying abilities have meant that harvesting and wine making on Welgemeend have always been a family affair. Ursula's contribution to the quality of the wines has been considerable. The couple's pioneering work in small-scale, intensive wine making has already influenced many others.

The wines are fresh and lively with soft tannins, and show a capacity for development with further maturation in the bottle.

Wine is for sale on the Estate on Saturday mornings.

Welgemeend

ROBERTSON

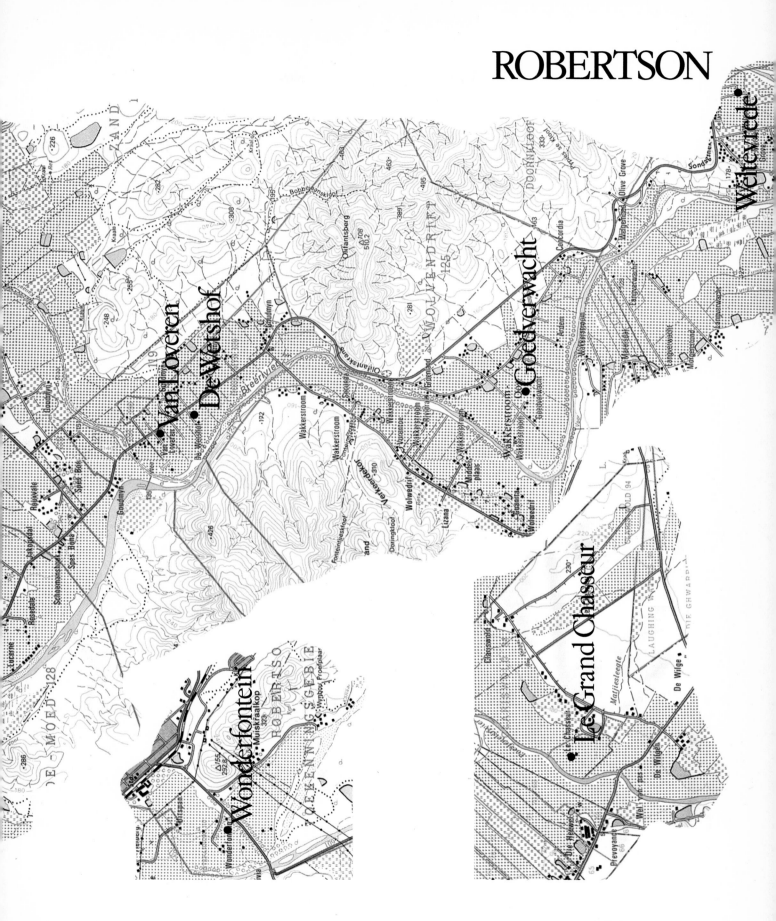

Bon Courage

The Bon Courage Estate is situated on a subdivided portion of the historic Goedemoed farm. For almost 150 years the Estate had been a centre of vine growing and wine making, until these activities were interrupted during the 1920s. At that time Bon Courage was bought by the Dutch Reformed Church in Robertson as a training school for farm managers, to help provide work for unemployed sons of local farmers. This period coincided with the ostrich boom and for a few years ostriches instead of vines were raised on the farm.

All Bon Courage's vineyards were removed by the trainee farmers and the cellar was used as a storeroom. Then in 1926 the scheme was declared bankrupt, and the farm was sold the following year to Willie Bruwer, who planted Steen and Muscadel vines and made wine from these varieties for the KWV. These were difficult years, and Willie, like most of his neighbouring farmers, was only just able to survive. The South African wine market was very small and the major export sweet wine markets that the KWV was later to develop were unknown.

In the early thirties Dr. Niehaus, a wine expert at the KWV, was developing techniques to make sherry on South African farms. In 1934 he persuaded Willie Bruwer to ferment Steen dry and make sherry with yeast found on the grapes in the vineyards.

This was the first time that commercial quantities of wine had been fermented dry in the Bon Courage (previously Goedemoed) cellar. The first owners of the 156-hectare Goedemoed property on the Breede River south of Robertson, Dirk Beukman and Jacobus van Zyl, planted vineyards, built a cellar and made sweet wine by adding brandy to their unfermented grape juice. This wine was transported to Cape Town by oxwagon once a year, for sale.

In those pioneering years, the Breede River had a seasonal flow and the vines often had to survive several months with insufficient water. The construction of the Brandvlei Dam near Worcester and the canalisation of the water supply to farmers downstream changed the style and scope of agriculture after the 1920s. It enabled Willie Bruwer to use flood irrigation from the canal on the alluvial deep-drained soils near to the banks of the river. The rest of the farm was used for grazing sheep.

The Bon Courage portion of the historic Goedemoed farm has the old manor house.

André Bruwer, the Estate's vintner.

A change of style in Bon Courage's wine production came with the arrival of Willie's son André on the farm in 1965. The market for sherry was decreasing and the consumption of dry and semi-sweet table wines was beginning to take the emphasis away from fortified and rebate wines in cellar planning. A major Stellenbosch wine merchant asked André to install cold fermentation techniques in the Bon Courage cellar and André made his first flowery, dry wines in 1968.

At that time the vineyards were still dominated by the Steen and Muscadel cultivars. André planted St. Émilion to extend his capacity of dry table wines and began to use sprinkler irrigation on his higher Karoo soils,

Bon Courage's three types of soil provide a healthy growing medium for vines.

which allowed him to extend the vineyards on the farm. Today half the Bon Courage vineyards are planted with Steen and Muscadel, and the rest with Colombard, Clairette blanche, Kerner, Bukettraube and Sauvignon blanc, all white varieties that are used for dry table wines.

André inherited 116 hectares of vineyard from his father and has since bought an extra 40 hectares, also part of the original Goedemoed property,

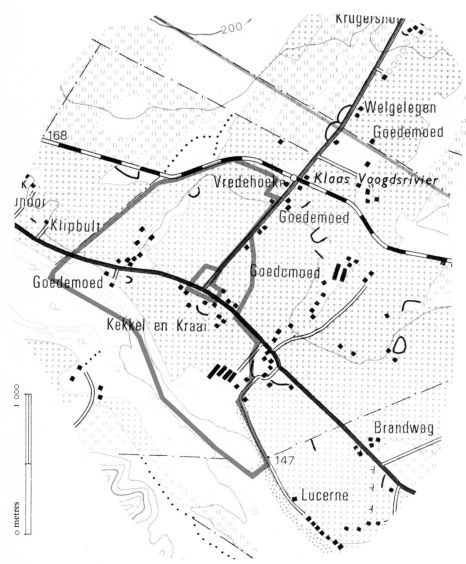

Pruning in the vineyards in winter.

which are now amalgamated with the Estate. He finds he has three distinctly different types of soil, each with different analyses. The two branches of the Klaas Voogds River run through the property, and they have deposited a strip of alluvial soil that probably originated in the Langeberg mountain range. This high pH soil shows a lot of promise for the production of quality wines. It adjoins a strip of sandy alluvium, the typical Breede River soil, on the banks of the river. "My better vineyards are on the rich Karoo soils higher on the farm, away from the river," says André. "I get more flavour from the grapes grown there. I've changed the whole farm over to drip irrigation and I can see the improved quality in the healthier vines and the quality of the wines in the tank."

André picks his cultivars destined for fermentation at 19-20 ° Balling as early as possible in the day. His crusher begins to squeeze grapes at 5 a.m., and the juice extracted from grapes brought in before 11 a.m. is kept separate from the must made from afternoon-picked grapes. He only gives skin contact to the juice of the cold grapes picked in the early morning. The juice is allowed to soak with the skins for 4-5 hours, then is drained off and fermented at about 14 °C until dry. The musts produced from grapes picked in the afternoon are fermented separately, also at 14 °C and also until they are dry. "Though we have a south-east breeze blow through here almost every summer afternoon from about three p.m., the afternoon temperature won't drop fast enough and we benefit mostly from the cool summer nights. The juice crushed early in the morning is green, and all of the afternoon juices are brownish."

Bon Courage vineyards produce fine quality Muscadel. André picks these grapes at around 25-26 ° Balling and makes a lighter style of wine than some of the other Muscadel producers in the area. The wine has a fresh, lively flavour.

Wine is sold on the Estate.

De Wetshof

Elevated maximum temperatures in summer and low natural rainfall make the production of high quality table wine in Robertson a daunting task, but there are three powerful forces in the De Wetshof mix that have lifted this Estate into the leading group of South African wine producers. The fertile, lime-rich soils and an unlimited supply of water in all seasons have combined with the driving ambition of Danie de Wet, backed by the steadfast support of his father, Johann.

Originally planted with Chenin blanc, Cape Riesling and other standard varieties, De Wetshof will soon have only premium varieties.

The boundless energy of the De Wetshof soil is probably the major building block in the development of the Estate. Sandy alluvial soil and the adjacent heavy Kogmanskloof alluvial soil, dominated by calcareous red soil with the Robertson area's characteristic outcrops of limestone, all share high pH levels and an outstanding ability to grow plants. In contrast with the relatively low fertility of the sandy and sandy loam soils of some of the vine-growing areas closer to the coast — which is believed to contribute to the quality of wines produced under those conditions — Danie and Johann believe that the outstanding health of De Wetshof's vines plays an important part in the quality of the juice obtained from the grapes.

The fertility of the soil plays a major part in the history of the farm and the surrounding area. De Wetshof was originally part of a considerably larger property, Goudmyn, which extended along the northern bank of the Breede River, near where the stream funnels into a sweeping curve and cuts through the mountains into the Bonnievale Valley. Goudmyn was roughly centred on the junction of the Kogmanskloof and the Breede Rivers and the reasonable supply of water and the rich soils combined to produce a stock-grazing property of outstanding merit.

The first settlers in this area obtained large parcels of land that were rented from the Dutch East India Company for 24 stuivers per year. When the English arrived, a stuiver was interpreted as a halfpenny. The land along the rivers and streams was farmed more intensively than the hills, and when the British revived the system of permanent grants of land the riverside portions were smaller than the open or mountainous farms. Goudmyn, one of the first permanent grants in the Robertson Valley after the introduction of the new system, became the property of Pieter Rossouw in 1815. The name indicates that his property had already shown itself to be the most valuable farm in the area.

The final decade of the nineteenth century saw ostrich farming spread from the birth-place of the industry around Oudtshoorn to other parts of the country where warm temperatures, low rainfall and relatively low humidity allowed these desert birds to thrive. Robertson became a major

Originating in the red Karoo hills, De Wetshof's calcareous soils have been deposited by alluvial action.

The vigorous growth of De Wetshof vines assists the development of noble rot in humid summers.

Danie and Johann de Wet, pioneers in top quality white wines.

The Estate's modern fermentation and maturation cellar.

production centre during the final boom period. In 1913 Goudmyn was bought by Pieter Potgieter, an Oudtshoorn ostrich farmer, who left a portion of the farm to each of his nine children.

In 1900 a major dam for an irrigation scheme on the Breede River was designed and construction begun by the government. When the water from the Brandvlei Dam became available to farmers downstream along the river, vineyards were planted and the first cellars were built. When Johann de Wet bought one of the sub-divided portions of Goudmyn in 1952, he found he had vineyards of varieties chosen and cultivated to produce large volumes of low alcohol rebate wine for brandy production. Johann named his new farm De Wetshof and planted his first Chenin blanc and Cape Riesling vines, to produce table wines in addition to traditional dessert Muscadel.

Johann de Wet is descended from a long line of Robertson De Wets and grew up on his father's property on the other side of the valley. He chose to buy a portion of Goudmyn to obtain the benefit of the dynamic soil.

Though summer days have high peak temperatures, Robertson benefits from a prevailing southerly, on-shore wind that brings relief to parts of the interior during most afternoons from January to April. This wind is also the major contributing factor to the coldness of summer nights in this area, lowering the average temperatures and providing a more stable medium for photosynthesis.

Johann's pioneering table wines were made in a simple cellar, designed for sweet wine production, ten years before the first refrigeration plant was installed in a Robertson winery, while Danie was still at school. Encouraged to study oenology, Danie studied for three years at Geisenheim, the German wine institute, learning the sophisticated cellar practice that would be required to make light, fresh table wines in the warm environment of Robertson. Danie's ambitions centred around the production of classic styles of wine, and on returning to the family Estate persuaded his father to install a cooling plant to reduce the temperature of the fermenting musts and make richly flavoured, delicate wines in the German style. The material that Danie began to work with, vineyards of Chenin blanc, Cape Riesling, Muscadel and bulk-production varieties, did not provide sufficient scope, and the wines were insufficiently rich and impressive to satisfy the demands of Danie's standards. Influenced by his German experience and instilled with memories of the luscious flavours of the classic sweet and semi-sweet wines of Germany, Danie encouraged his father to plant small blocks of every quality German variety that could be obtained. Soon De Wetshof had rows of Rhine Riesling, Kerner, Bukettraube, Rülander (Pinot gris) and Gewürztraminer vines. These were to be used as propagation material for the development of future vineyards.

The mid '70s saw the success of the De Wetshof father and son team at

wine shows and on the market with their dry wines made from Cape Riesling and Chenin blanc. After consultation with Dr. Julius Laszlo of the Bergkelder, the De Wets decided to concentrate on the production of dry wines and to plant vineyards of Chardonnay and Sauvignon blanc that would provide grapes with the broad varietal flavour required to make classic table wines. From the experimental vineyard, only Rhine Riesling was chosen for propagation, becoming the third major variety producing dry white wine for the De Wetshof label. The Riesling and a limited quantity of Steen remained on the market until the initial commercial quantities of the three premier varieties were ready. The 1979 Sauvignon Blanc and a blend of the '80 and '81 Chardonnay wines were the initial releases, and they established new price platforms for young dry white wines made in South Africa.

After several seasons' experience with these classic varieties, Johann and Danie isolated a number of factors that influence the condition of the grapes arriving at the De Wetshof cellar and the subsequent quality of the wine. All three of the chosen classic white varieties, together with Chenin blanc, ripen during the earliest part of the harvesting season and are ready to be harvested at the same time, causing a bottle-neck in the picking programme. This probably indicates that De Wetshof will switch to mechanical harvesting when fur-

New presses will allow Chardonnay and Sauvignon blanc juice to be given extensive periods of skin contact.

Grapes are picked in baskets, then transferred to small 2 tonne trailers for transport to the cellar.

ther vineyards of the early varieties provide too great a task for hand labour. Sauvignon blanc and Rhine Riesling bear similar sized quantities of grapes, but the Sauvignon blanc has larger bunches and is much easier to harvest than the German variety.

Chardonnay has small bunches and produces a light crop of about seven tonnes per hectare on De Wetshof. In all, Danie receives about 80 per cent of his grapes in the first two weeks of the season, and the rest of the crop (Colombard, Cape Riesling etc.) trail in during the following three weeks. The final week of harvesting in the cooler and more moist years sees the arrival of the Weisser Riesling, Sauvignon blanc and Chenin blanc grapes that have been infected with *Botrytis cinerea*. These are left to concentrate their sweetness for the production of De Wetshof's Edeloes Noble Late Harvest.

The first grapes to be picked are normally Chardonnay, which has shown a tendency to over-ripen and an ability to attain very high concentrations of sugar, retaining reasonably high acidity. The heat of a warm Robertson summer and the proximity of the other early-ripening varieties make the harvesting of Chardonnay at its critical point a delicate exercise. During the first three seasons of

Chardonnay production, Danie used only the free-run juice and a limited quantity of pressed juice, but has now installed tank presses that enable him to leave the juice with the skins and stems from the bunches to get extra flavour. The plan is to pick the Chardonnay at a pH of around 3,2, when the grapes are fully ripe, but with a fresh acid taste. After the juice has had sufficient contact with the grape husks, the skins are thoroughly pressed to extract all the remaining liquid for the must, which is then fermented at relatively warm temperatures (17-18 °C). After the must has fermented dry, the new wine is run through a centrifuge to clean it thoroughly before it is placed in oak casks to mature and get extra character from the wood. Danie believes that Chardonnay needs at least six weeks contact with new oak. "Even longer, if it has plenty of alcohol and acidity," he says. Chardonnay grapes need to be fully ripe for the wines to develop the characteristically broad, mouth-filling style, and these deep, full-flavoured wines, with tannin extracted from the grape skins, require moderation

Most of De Wetshof's premium variety grapes are crushed in the Estate cellar during the first two weeks of harvesting.

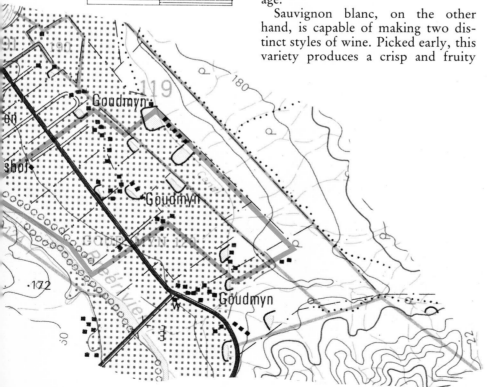

through the vanilla flavours obtained from the new oak and the benefits of age.

Sauvignon blanc, on the other hand, is capable of making two distinct styles of wine. Picked early, this variety produces a crisp and fruity wine with a distinct 'grassy' aroma and taste. Allowed to ripen fully, it gives a broad, full-bodied wine, with almost a muscat aroma, that gains an added dimension from wood ageing. The initial De Wetshof Sauvignon Blanc wines were made in the latter style, with yellow-ripe grapes that have produced fairly high alcohol wines.

The grape skins are left in the juice for flavour extraction, then given a heavy pressing to get all the available liquid for the must. Fermentation follows at a cool temperature (15 °C) and all the sugar is removed from the must. Over a period of about three weeks the new wine is cleaned and matured in oak, normally for less than two months, before filtration and bottling. To reduce the hazards of a period of high temperature during the first weeks of harvesting, causing two of the most important varieties to ripen together and possibly producing too many full-bodied wines, the De Wetshof programme now includes an additional Sauvignon Blanc style made from early-ripe grapes. This wine has a light, fresh 'grassy' flavour and can be blended with fuller wines,

The De Wetshof cellar and vineyards are in the left foreground and run down to the banks of the Breede.

if required, to produce a more evenly balanced product.

The German variety, Rhine Riesling, is used to make lighter, more fragrant wines. Although it presents difficulties during wine making because of its small, light bunches of fleshy grapes that do not easily release their juice, Rhine Riesling has made some of the most successful De Wetshof wines in recent years. The crushing procedure requires the stems of the bunches to be retained, providing more space within the mash for the juice to drain off freely, without any deliberate period of skin contact. The juice is cleaned and fermented at moderate temperatures (16-17 °C), then, when dry, the wine is separated from all solid matter and retained in closed steel tanks to be made ready for bottling without maturation in oak.

The three wines made from the classic white varieties are given a period of bottle maturation before sale. De Wetshof makes a fourth style of wine, thick and honey-sweet, from the late-picked 'noble rot' grapes of selected varieties in those seasons when the prevailing weather encourages the development of the beneficial fungus. The grapes are allowed to remain on the vine, losing moisture and concentrating sweetness and acidity until the sugar content passes 40 ° Balling. The juice is chilled, cleaned through the centrifuge, then inoculated with yeast to begin fermentation, which proceeds until halted by the combined concentration of increasing alcohol and decreasing sugar. The wine is again chilled and centrifuge-cleaned — a practice that Danie uses for all his wines — before maturation. Only occasional vintages of Edeloes, as Danie calls the De Wetshof Noble Late Harvest, are matured in wood. The richer, more full-bodied sweet vintages are given an extra dimension of oak character before bottling and a further period of age before release. Each wine is judged on its merits, with, for instance, the soft, round, bottle-aged 1981 Edeloes being released several years before the harder 1980 wine, which had the additional benefit of wood contact.

The De Wetshof father and son team deserve the high regard of both consumer and producer for the whole-hearted manner in which they have planted extensive vineyards of varieties that, regardless of their international standing, had been largely ignored in South Africa.

Wine is not for sale on the Estate.

De Wetshof lies on level ground with calcareous soils and alluvial river soils in the narrow end of the Robertson Valley.

Excelsior

The central section of the valley between Robertson and Ashton, with rich alluvial and calcareous soils, originally cleared to graze cattle, is the location of the Excelsior Estate, sandwiched between Zandvliet, De Wetshof and Rietvallei.

Jacobus Stephanus de Wet, second son of a stock-farming family from the Worcester district, bought a large piece of ground, called Zandvliet, on the Kogmanskloof River around 1860 and grazed sheep and cattle, bred and raised horses and had a small vineyard. He divided his property between three sons, with the third son Jacobus Stephanus (Kowie) inheriting the portion now known as Excelsior. Kowie took advantage of the growth of the ostrich industry and became the largest ostrich breeder in the Robertson area. When the ostrich feather market collapsed Kowie and his son Oscar concentrated on the breeding of sheep, cattle and horses, and began to develop the farm's vineyards. In 1967 Oscar's son Stephen finished his studies and began farming Excelsior, and in 1970 his brother Freddie joined the team.

During the '67 and '68 seasons Stephen used mule carts designed and built by his father to fit between the narrow rows of the old Muscadel vineyards to bring his grapes to the cellar. Since then the brothers have instituted a programme designed to lift Excelsior to the forefront of both wine producers and race-horse breeders. They have divided the farm's major responsibilities between them, with Stephen in charge of the horses and the cellar and Freddie controlling the extensive irrigation and crop-spraying systems.

Excelsior has mostly level ground, with some gentle southerly slopes running down to the Kogmanskloof River. There are three distinctly different types of soil and vineyards are grown in each of them. The rich, black alluvial soils originating in the valley on the other side of Montagu and the red alluvial soils beside them are found alongside the Kogmanskloof River. Away from the river, on slightly elevated ground, the most common soil type is calcareous. This latter soil is much less fertile than that found near the river, and the yields of vineyards grown in the lime-rich ground are restricted, but the quality of the wines made from grapes grown on these soils is far superior.

Though sharing the Little Karoo's climatic features of hot summer days and low rainfall, Robertson benefits from the cooling effect of the south-east coastal winds that lower temperatures and lengthen the ripening season of grapes. A weather monitoring station on the farm Prospect, adjoining Excelsior, has shown that January has the highest maximum temperatures, and greatly reduced average temperatures for March and April. Robertson is famous for the summer mists that rise in the early morning and last until about 11 a.m.

Excelsior's magnificent homestead was built by Jacobus de Wet, Robertson's leading ostrich farmer.

Robertson's high humidity in summer is demonstrated by the frequent morning mists.

Excelsior's vineyards are grown on mostly level soils running down to the river.

Every winter sees the removal of some old vineyards for the planting of better-quality vines.

Freddie (left) and Stephen de Wet, who produce wine and fast racehorses on Excelsior.

When Excelsior's Muscadel vines are pruned to restrict the size of their crop they are capable of reaching sugar concentrations in excess of 24° Balling, while retaining a significant amount of acidity. Both red and white Muscadel varieties ripen together, as one is simply a mutation of the other. After the grapes are crushed the skins (pomace) are left mixed with the juice for between 24 to 48 hours to extract the maximum of the desired colour and flavour before separation. The mixture is not cooled and the vineyard yeasts on the grape skins begin to ferment the sugar in the grape juice. Stephen believes that this 'bubbly' stage is essential in the production of a good Muscadel. The mash when ready goes into the press, where the liquid content is freely run off, and then the skins are pressed to obtain the last few extra drops of richly-flavoured liquid. The liquid is mixed with enough pure wine spirit to halt fermentation. Solid particles in this new sweet wine settle to the bottom of the tank and are removed by racking the wine, which is then allowed to mature inside a closed vessel for several months.

Though Excelsior is best known for the quality of its Muscadel, the proportion of sweet wine to dry white wine made in the cellar has swung in favour of the table wines, and Excelsior now produces twice as much dry white wine as Muscadel. The Estate has Chenin blanc, Colombard and Trebbiano (Ugni blanc) vineyards that ripen before and immediately after the Muscadel and Muscat d'Alexandrie. When the grapes for fermentation are crushed they are allowed to stand for at least two hours, with skins and juice mixed, in the tank press. When the grapes have been picked in the early morning, and the juice is comparatively cold, Stephen will leave the juice in contact with the skins for at least four hours. The clear juice is chilled, inoculated with yeast and fermented cold (14 °C) until dry.

The production of dry table wines is expected to diversify over the next few years, with high-quality, low-production white varieties being planted and the possible introduction of quality red varieties into the vineyards.

At present, wine is not for sale on the Estate.

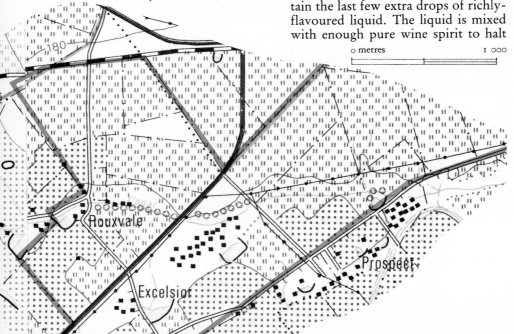

Goedverwacht

Looking across the vineyards toward the source of the morning sun, beyond Bonnievale.

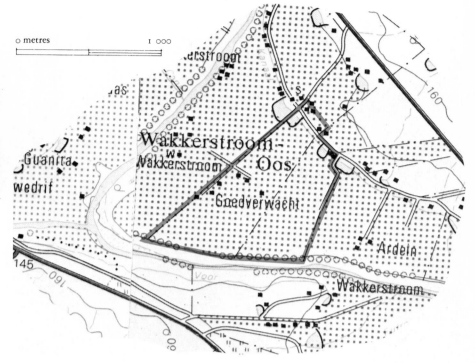

Enclosed within a sweeping bend of the Breede River, soon after it enters the Bonnievale Valley, Goedverwacht contains 61 hectares of rich, level alluvial soils. Gawie du Toit, owner of Goedverwacht, believes these soils to be capable of pushing the sugar content of his grapes to the highest concentrations in the Bonnievale area. His grapes are the first to reach ripeness in the valley and he believes the low eastern horizon and long hours of sunlight to be responsible.

The Estate is part of Wolvendrift, one of the original loan farms in the Robertson and Bonnievale area, 4 500 hectares of which were granted to Pieter Marais in 1843. Because all the land within the wide bend of the river is flat, and because the farm is downstream from the junction of the Kogmanskloof and Breede Rivers, providing two sources of water for a more reliable supply, a large portion of Wolvendrift was irrigated, even before the construction of dams on the river. Goedverwacht was divided from Wolvendrift during the nineteenth century.

Goedverwacht has seen a long history of intensive farming under irrigation, and down through many generations the water from the Kogmanskloof River, high in salt content, that flowed directly into the Sanddrift Canal has given the Estate's soils a very high saline content. However, vine roots are not at home growing in ground with such a high degree of salt.

"When my father bought this farm in 1961," says Gawie du Toit, "the vineyards were in a sorry state. Old vines weren't producing. Young vines wouldn't grow. My first important project was to readjust the soils to allow me to make a living."

But the problem of salt content was not isolated to the Goedverwacht farm, and a syphon was installed to divert the Kogmanskloof and keep its water out of the irrigation channel. This improved the water supply analysis, but Goedverwacht soils contained salt deposited during more than 150 years of irrigation with Kogmanskloof water.

Gawie progressively took out old

Goedverwacht's saline soils had to be provided with better drainage before top varieties could do well.

Goedverwacht's vineyards are the earliest to ripen in the Bonnievale area.

With rainfall of less than 200 mm per year, irrigation is necessary for the growth of the vines.

vineyards and installed deep drains to provide permanent efficient draining. Gradually the excess salinity was washed out of the soil by the regular irrigation programme. Gypsum was thoroughly mixed with the deep-ploughed soil, as Gawie believes it improves the porosity of the soil and provides better drainage.

At the time of the Du Toit purchase, Goedverwacht was still using the method of irrigation that had been the practice since the farm was first settled. Drainage furrows were dug from the vineyards to the sluice gates on the river water channel. Vineyards and orchards were given periodic flooding to provide the necessary water for the vine roots. First switching to sprinklers, then to drip irrigation, Gawie has made more efficient use of the water supply and the health and production of the vines have improved.

Gawie du Toit is a relative newcomer to Bonnievale and vine growing. Born on a wheat farm in Moorreesburg, he studied Civil Engineering at Stellenbosch University and worked for the Orange Free State Provincial Administration for six years as an engineer. Then, to satisfy Gawie's farming ambitions, his father, Thys, bought Goedverwacht and financed the initial development.

At that time, the vineyards contained Steen, Muscadel, Palomino, Cinsaut, St. Émilion and various bulk-production varieties, almost all of which were grafted on to Jacquez rootstock. Gawie had choice selections of Chenin blanc, red and white Muscadel, Colombard, Raisin blanc and St. Émilion grafted on to better root material and planted these in his newly-drained vineyards. Since then, the Chenin blanc and Colombard vineyards have been enlarged and the Estate cellar produces four times as much dry table wine as fortified dessert wine.

Gawie du Toit believes that the quality of the grapes harvested determines the quality of the wine produced, and the Estate has been planned to convey the freshly-cut bunches to the cellar in the best possible condition. Gawie has designed and constructed small wagons that can be towed by tractor between the rows. Once they have been filled the wagons transport their two-tonne loads of grapes to the cellar, located in the centre of the Estate, and are crushed as soon as possible after picking. The grapes have minimum contact with the air and so the chance of oxidation of their juice is reduced.

Gawie has provided his sons with the background knowledge necessary for the production of high quality table wines. Thys, the elder son, is a graduate of Stellenbosch University, with Jan, the younger son, a cellar technology student at Elsenburg Agricultural College.

Wine is not for sale on the Estate.

Le Grand Chasseur

Colombard planted by mistake on Le Grand Chasseur became a success and is now a popular variety.

Most Robertson farms have large areas of veld. In winter, the sheep also graze in the vineyards.

The area today known as Le Chasseur lies along the south-eastern bank of the Breede River before it reaches the town of Robertson. The name Le Chasseur was first given to the area by Jan David Storm Jnr., when he was granted the property in 1841.

A part of the original property was bought by the brothers Jacobus and Hendrik de Wet in 1881. Like the previous owners, they grazed sheep and cattle on the farm with a limited amount of cultivation restricted to the land beside the river. This agriculture was further limited to those months of the year when the river flowed, as the Robertson district receives only about 200 mm per year, almost all of which falls in the winter months.

Wouter Johannes de Wet owned a large part of Le Chasseur at the time of the construction of the Brandvlei Dam on the Breede River, near Worcester. The Dam changed the character of farming along the course of the river. Water canals were provided to allow year-long cultivation on the rich soils of the Worcester, Robertson and Bonnievale Valleys.

Wouter de Wet's son, Albertus, was one of the first five students to obtain a B.Sc. (Agric.) at Stellenbosch University. In 1929 Albertus became the owner of a 2 000-hectare portion of Le Chasseur that has since become known as Le Grand Chasseur. He grew lucerne and vines on the newly irrigated ground and grazed merinos on the veld slopes. Albertus de Wet built the first farm school for the children of farm labourers in the Robertson area on Le Grand Chasseur in 1937. In later years the school was increased and today there are seven classrooms and about 250 pupils.

Until about 1950 the vineyards were used mainly to produce sultanas and raisins. Some sweet wines were being made in the cellar but the real wine making began from about this period, when a Hanepoot wine was made for the KWV to export to Canada. The must was allowed to ferment until a little more than half of the sugar in the grape juice had been removed (10 ° Balling). At this point wine spirit was added to prevent further fermentation, and a type of fortified semi-sweet wine was made.

Traditionally brandy had been an important sector of the liquor market and sales of commercial brandies increased during this period. Though Robertson farmers were keen to make dry wines for brandy production, they were told by the liquor industry authorities that their area was unsuited to the production of brandy and that they should remain sweet wine and sultana producers. Under pressure from the grape growers, the KWV installed the first distillery in Robertson around 1945. This became the biggest distilling house south of the equator and Robertson became the largest brandy production area in South Africa.

Albertus was succeeded by his son, Wouter de Vos de Wet, in 1950. At that time, the farm was allowed one cubic yard of water per second. When the allocation was increased to 2,55 cubic yards, Wouter was able to increase the area under vines to around 150 hectares, all under irrigation. Planning additional vineyards in 1956, Wouter ordered a quantity of St. Émilion (Ugni blanc) vines and by accident launched a new grape cultivar into the South African wine industry. When the first bunches appeared on the new vines, Wouter discovered that he had been given the wrong cuttings and had planted Colombard. He contacted Beyers Henning of the KWV, hoping to learn something about the suitability of Colombard for the production of re-

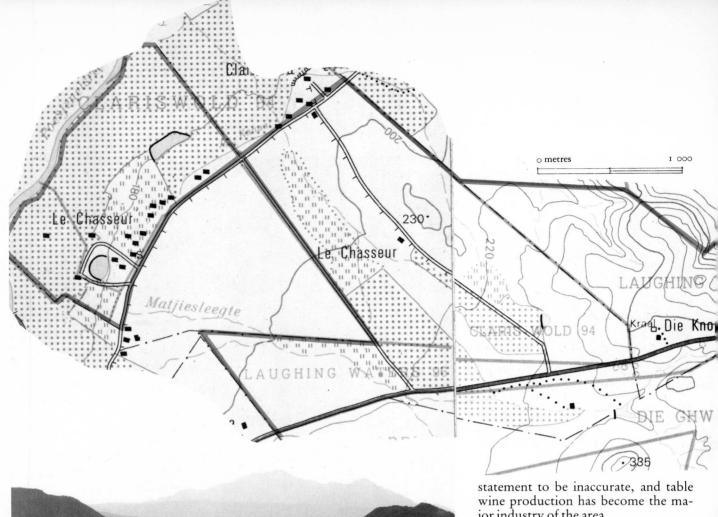

Looking over the vineyards beside the Breede River toward the Langeberg Mountains.

bate wine, but was encouraged to experiment with the making of sweet and semi-sweet wines.

The period around 1960 was a major turning point for Le Grand Chasseur and Wouter de Wet. Not only did the discovery of the interesting wines that could be made from a new variety bring attention to the farm, but it caused the style of wine making to change radically.

By 1968 Wouter had a large area of Colombard producing a substantial quantity of wines with individuality and considerable merit, and was encouraged by a wine merchant to install cold fermentation equipment, to make dry and semi-sweet wines. The success of these wines caused intensive planting of Colombard in the Robertson area.

After proving their capability of making superior rebate wine, Robertson wine producers became interested in the table wine market, but were discouraged by an official statement that Robertson's climate would prevent the production of quality dry table wines. Wouter's success with Colombard and the subsequent experimentation by others proved this statement to be inaccurate, and table wine production has become the major industry of the area.

Since then, Wouter has planted Cape Riesling, Clairette blanche, Chenin blanc and St. Émilion. Today all the vineyards contain white cultivars and have been equipped with drip irrigation. The cellar has swung almost wholly to dry table wine production. About ten per cent of the crop comes from Muscadel vines and is used for the farm's only remaining sweet wine.

The grapes for the dry wine are picked fully ripe at 20-21 ° Balling. After crushing, the juice is chilled to 13 °C and held at that temperature overnight to allow any solid matter to settle. Yeast is then added and the temperature is held at 15-16 °C and adjusted to ensure a regular rate of fermentation.

When the sugar content of the juice has reached around 1 ° Balling, the temperature of the must is allowed to rise, to ensure that it will ferment dry.

In 1979 Wouter's son, also named Albertus, finished his university studies, becoming the third generation to obtain a B.Sc. (Agric.) and joined his father in a partnership on the farm.

Wine is not yet for sale on the Estate.

Mon Don

In 1944 Hannetjie Marais inherited 50 hectares of the historic Goedemoed farm in the Robertson district. The farm had a cellar without a roof, unpromising vineyards of Steen, Palomino, Cinsaut and many other varieties, and there was a peach orchard. Though the soil was very fertile and grapes ripened easily, Hannetjie's portion did not have enough of Goedemoed's original water rights to water the whole property, and a part of the farm was unplanted.

Hannetjie's husband, George Marais, was the Chief Librarian at the Carnegie Library at Stellenbosch University, and the couple lived in Stellenbosch. The property in Robertson was run by foremen for almost twenty years. Then in 1962 the Marais's son Pierre moved on to the farm, built a house and began to redevelop the vineyards. Because he had grown up in the town of Stellenbosch with little experience of his mother's vineyards, he learnt viticulture and wine making at Elsenburg Agricultural College in preparation for turning the farm into a commercial enterprise.

When Pierre moved to the farm, Hannetjie sold him the property on terms that could be described as a gift, and he has since commemorated her generosity in the name of the Estate, Mon Don, 'my gift'.

Derelict vineyards and insufficient water provided sizeable handicaps for a recently graduated viticultural student. The farm had three hectares of peaches in good condition and, until some replanted vineyards were producing, the fruit trees provided the major source of income.

Hannetjie and George Marais approached a Stellenbosch merchant and obtained an outlet for Pierre's as yet unmade wines. In 1963 Pierre made sweet wines from the existing vineyards of Steen and dry wine from the other varieties. In 1964 he made flor

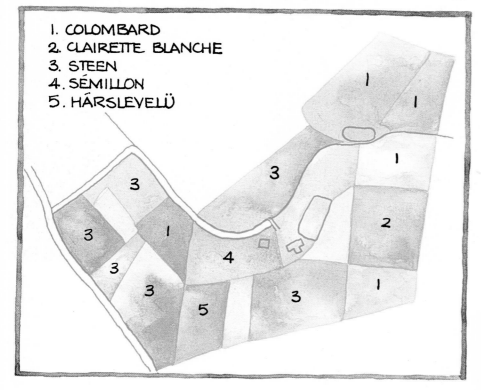

1. COLOMBARD
2. CLAIRETTE BLANCHE
3. STEEN
4. SÉMILLON
5. HÁRSLEVELÜ

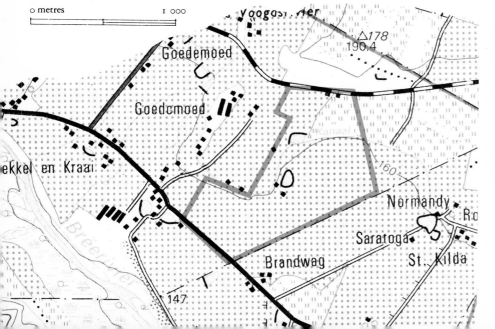

Pierre Marais produces only dry white wines in his temperature-controlled cellar.

Mon Don's calcareous soils help grapes to ripen fully without strain.

sherry and sweet sherry for the KWV, but the market began to deteriorate and soon he was producing only dry wines for merchant sale, which is still the case today.

The soil of Mon Don's 77 hectares varies greatly over short distances. Though its chief characteristic is the high calcium content from the outcrops of lime that are common to the whole district, its texture varies from limestone rock to deep sand to gravelly loam to a heavy soil with a clay base. Pierre has broken up the limestone outcrops typical of this area and mixed these thoroughly with neighbouring soils to provide consistency within the vineyards.

To redevelop the vineyards, Pierre has planted new blocks of Steen, Clairette blanche and St. Emilion, and has added Colombard. More recently he has taken out the peaches and planted Hárslevelü in the soil made available.

"We have a unique weather pattern," says Pierre. "We have high light intensity in summer. The days get very hot but the nights are relatively cool from February to April. And we seldom have more than two or three days of hot weather running. The south-easter begins to blow and the thermometer falls back to a more moderate figure. This gives our wines freshness. The grapes ripen fully, but they don't strain."

The south-easterly wind blows from the Antarctic over the southwest Cape. It blows up the Breede River and from Agulhas bringing moisture that condenses to give Robertson a regular pattern of morning summer mists. The cool mornings last until eleven or twelve o'clock, and are followed by hot weather during the afternoon. The temperature drops quickly in the early evening.

Pierre would prefer to pick all his grapes before midday each day. "It just isn't possible. When a block of vines is ready, the grapes have to come straight into the cellar. Then I keep the hot afternoon juice apart from the cool morning juice and ferment them separately. The cooler juice makes much better wine.

"I think that Steen is full-ripe at around 22° Balling, but I have to start picking before it reaches that figure and finish after. My other varieties are ripe and ready with less sugar and they make lighter wines. This farm has the ability to reach very high sugars, and I think that it's best suited to making full-bodied wines. But the demand, at the moment, is for light, fresh wine, so that's what I'm making."

Pierre allows the crushed skins to soak in the juice for a period of time, then cools it rapidly to 13 °C and leaves it to settle out overnight. The clear juice is separated from the lees, which is filtered to obtain the rest of the juice. The wines have pleasing aroma and are easy to drink when fresh.

At date of publication, wine is not yet available for sale on the Estate.

Mont Blois

Picture a tiny valley, less than one kilometre wide, green with trellised vines, backed by a range of mountains, rising in an almost sheer wall to peaks over 1 500 metres above the valley floor. This is Mont Blois, one of the most spectacular and attractive farms in South Africa, situated a few kilometres to the north-east of Robertson.

Because Robertson gets little rain in winter, and virtually none at all during the rest of the year, every vine on Mont Blois is irrigated. "You can't farm here without irrigation," says Ernst Bruwer, owner and vintner of Mont Blois Estate. "Without extra water, every vine on the farm would die."

Robertson is regarded as the finest area in the land for the growing of Muscadel vines and the making of sweet fortified red and white Muscadel wines. The constitution of the red Karoo soil, with its moderate fertility and high calcium content, allows the vines to ripen their grapes to high sugar levels with considerable natural acidity. The combination of richly-flavoured juice with balanced sweetness and acidity provides the basis for one of the world's top fortified wines.

The Mont Blois Estate comprises three separate sections, each representing a variation on the standard soil type of this region. The home farm is located on the deep alluvial soils of the valley floor. La Fontaine, situated at the broad entrance to the valley, has traces of Robertson's famous calcareous soil, and on the other side of the Breede River Valley a portion of the historic Goedemoed farm has sandy loam soils to add variation to the crop coming in to the Estate cellar.

The towering Langeberg Mountains dominate the narrow valley.

Mont Blois is one of the top Muscadel producers in South Africa, and has been particularly successful with white Muscadel. Ernst Bruwer has won numerous championships with this wine, which has been nationally distributed within South Africa since 1975. Since its first release, until the date of publication of this volume, every bottle has been certified Superior.

It is not known how long the Bruwers have farmed in Robertson. There is no doubt that they have been farmers in the Cape Province since the days of the Huguenot emigration. Although the first farms in the Robertson area date from around 1800 — some one hundred years later than those of Stellenbosch — the annual summer drought that dries the Breede River to a rock bed prevented anything but grazing in the Robertson and Bonnievale Valleys before the Brandvlei Dam was built in 1919.

The buildings are on the smallest of the three portions of the Estate.

The first Bruwer family came from Blois in France with the Huguenot refugees in 1688.

Since the early 1800s, however, irrigation was used to maintain year-round cultivation along the edge of the mountain range where Mont Blois is today. On the Estate is a dam that has been fed by a mountain spring throughout the year for at least a century, and there Ernst's ancestors grew fruit and vegetable crops during the late nineteenth century.

The farm's name is derived from a combination of Blois, the ancient French home of the Bruwer family, and the farm's adjacent mountain.

Mont Blois Estate contains over 3 500 hectares, of which 95 are used for the growing of vines, with the rest of the land either mountainside kloof or semi-barren veld, used for grazing stock. The Muscadel, Chenin blanc, Palomino and Hanepoot vineyards have been used by Ernst for the production of both fortified sweet wines and dry table wines. Though warm summers, bringing higher sugar levels in the grapes, encourage the use of a greater proportion of the juice for sweet wine production, both styles of wine are made during every vintage. In future, the pendulum will gradually swing in favour of dry and semi-sweet wines, and Weisser Riesling, Sauvignon blanc and Chardonnay vineyards are being planted on the Estate. In 1978 Ernst Bruwer's son, Pieter, joined the Mont Blois management team and has taken charge of the Estate cellar. Pieter is a graduate of Stellenbosch University, and spent three months of 1977 working on wine farms in Germany.

All the Mont Blois vineyards are equipped with drip irrigation, which allows a controlled and pre-set amount of water to be added to each vineyard, as required. Each vineyard receives either two or four litres of water per hour from the permanently installed system, which also provides a regular supply of fertiliser, dissolved within the liquid.

Pieter Bruwer harvests the Estate's Muscadel grapes fully ripe, when the skin of the berry has a slightly wrinkled appearance. After crushing, the skins of the crushed grapes are left mixed with the juice for maximum flavour extraction. After fermentation has begun, the skins are removed and pressed to get the last of the rich flavour, and enough wine spirit is added to the liquid juice to halt fermentation. The white Muscadel wine is stored in the Estate cellar for several months before transportation to Stellenbosch for bottling and further maturation.

Ernst Bruwer, producer of South Africa's best known white Muscadel wines.

The table wines are fermented dry in the Estate cellar, using controlled fermentation, and together with the other fortified wines are sold to a wine merchant for blending and sale under proprietary brand names.

Wine is not for sale on the Estate.

Rietvallei

Johnny Burger made his first dry table wines in 1982.

Looking across the Rietvallei vineyards toward Bonnievale.

For more than a century the farming lands along the Breede River and beyond were used by Stellenbosch and Paarl farmers for grazing. A successful farmer in the coastal areas could easily rent grazing and hunting land on the far side of the mountain. Rent on those properties was paid annually to the Dutch East India Company, and this practice was continued for a short period by the British after they took over the governing of the Cape Colony at the beginning of the nineteenth century.

The major hazards of grazing cattle and sheep on the frontier of the settlement were the tendency of lions and other predators to treat the settlers' livestock as fair game, and raids by 'Bushmen' who came out of the uncharted interior and carried off livestock and other spoils.

In 1717 Claas Voogt and a companion set out from a farm in the valley to catch a marauding band of 'Bushmen' and teach them a lesson. The intrepid pair followed the raiders into the foothills of the Langeberg Mountains, where they surprised an elephant. Claas was trampled to death by the enraged beast, and scarcely a bone was left in his remains. Two of the streams coming out of the Langeberg Mountains were named after this pioneer. The sources of the East and West Klaas Voogds Rivers are not far from the scene of Claas's death. They bring fresh water for most of the year down into the Robertson Valley and flow eventually with the Breede River within two hundred metres of one another on to the Bon Courage wine Estate.

Until 1820 Pieter Moller rented land on elevated ground in the centre of the valley. The two streams brought water for his cattle. Piet had applied for a land grant, but before it was granted he died, leaving the land to be registered as 'Rietvallei van de Klaas Voogdsrivier Wagendrift' in the name of his widow, Aletta Nel. In 1864 Jacobus François Burger bought

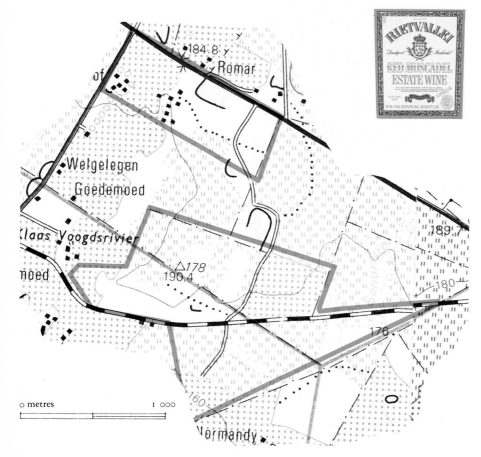

Rietvallei's soils are believed to have been brought from the Langeberg Mountains by the Klaas Voogds River.

Rietvallei and continued the tradition of grazing livestock. The farm had a small cellar, and a vineyard supplied grapes for the making of sweet wine.

Johannes, Jacobus's son, planted more vines and diverted water from the mountain streams to provide moisture for the vineyards. His first crop of 50 tonnes of grapes was turned into wine in the old cellar. When he expanded his cellar, the townspeople of Robertson came out to the farm to see how concrete was used. In 1908 Johannes planted a vineyard of red Muscadel next to the cellar, with vines and rows respectively one metre apart. This vineyard was to become one of the most famous dessert wine sources in South Africa. It was to establish Rietvallei's fame as a sweet wine Estate in the years following 1975, when the wines were first made available nationally.

"We had to work it by hand, the way they used to care for and pick all the vineyards in the old days," says Johnny Burger, Rietvallei's wine maker. "Part of the vineyard had a few other varieties mixed with the Muscadel and we removed this section in 1981.

"We then prepared the soil to take it back to its best and we've planted more Muscadel there, again closely

The Estate's famous old Muscadel vineyard has to be cultivated by hand because the vines are planted one metre apart. Grapes have to be carried out in baskets.

spaced, but we've widened the rows so that we can get a tractor in to cultivate and help during harvesting."

In 1948 Gideon van der Vyver bought a piece of the historic Goudmyn farm, and in the same year Jacobus Burger, Johannes's son, married Ria, Gideon's daughter, and moved to the Goudmyn property to live. There was no cellar on Goudmyn, so Jacobus became a founder member of a Robertson co-operative in that same year.

The Rietvallei cellar dated from before the turn of the century and was only suitable for dessert wine production and the market for sweet wines was declining. The cost of converting the old cellar to cold fermentation forced Jacobus to join another co-operative in 1971. He continued to make his Muscadel in the old cellar, but the rest of the crop was sent to the co-operative for fermenting into dry wines.

In 1973 Johnny Burger, Jacobus's son, took over the running of Goudmyn, and in 1975 he also took charge of Rietvallei. He continued the practice of making sweet Muscadel in the old cellar, with the rest of the crop being sent away for fermentation and eventual sale. In 1975 Jacobus arranged with the Bergkelder to bottle and distribute the farm's red Muscadel nationally.

As a major change of policy, in 1981 Johnny built a cellar large enough to ferment dry the juice of his red and white cultivars, as well as to make his traditional red Muscadel, and the emphasis of the farm is swinging toward table wine production.

Rietvallei's vineyards are on the crest and south-eastern slopes of the lower foothills of the Langeberg Mountains. The soils are rich, with a high concentration of lime, like most other Robertson farms, but have a remarkable ability to provide vines with the capacity to concentrate their sweetness and reach very high levels of sugar. On other Robertson Estates this has been shown to benefit classic table wine cultivars picked at lower sugar concentrations than Muscadel. It seems likely that Rietvallei will follow in their steps with top quality dry wines.

Johnny has vineyards of Steen, Raisin blanc, St. Émilion, Colombard, Sauvignon blanc, Shiraz and Pinotage, in addition to red Muscadel.

Wine is not for sale on the Estate.

Van Loveren

The two poplars were growing on Van Loveren when Hennie Retief arrived in 1937. Both the tree on the left and Hennie died in 1982.

"We've planted a lot of new varieties and I plan to make fresh, light wines from each of them," says Wynand Retief, wine maker at Van Loveren. "Our first Fernão Pires was picked a little late and we made a very full-bodied wine with it that won a gold medal. But I would prefer to make only lighter drinking wines, like our Hárslevelü, something that's pleasant in any season of the year."

The Van Loveren cellar stands on part of the historic Goudmyn farm on the Breede River. Goudmyn was one of the first farms granted in the Robertson district when it became the property of Pieter Roussow in 1815. During the years of the ostrich boom, Goudmyn was bought by a family called Potgieter who brought ostriches from Oudsthoorn, and grazed them on pasture beside the river. Mr. Potgieter's will divided Goudmyn into nine pieces, one for each of his children. One of the 28-hectare sections was sold to Jan Hofmeyr, who was then a fruit farmer on the Breede

A small trailer provides greater speed between vine and cellar.

River. Hofmeyr sold this property in 1937 to Nicolas Retief, a farmer of Klein Drakenstein. Retief bought the farm for his youngest son, Hennie, who moved on to the farm and married Jean van Zyl, a direct descendant of Guillaume van Zyl and Christina van Loveren. They renamed the property Van Loveren to honour Jean's forebear. Jean is the proud owner of a kist that Christina brought from Holland in 1692.

In 1964 Hennie bought Schoemanshoek, a small farm separated from Van Loveren by two intervening properties. Four years later the family bought another small piece of land alongside Schoemanshoek. The grapes from each section were crushed in the Van Loveren cellar. In 1979 another 39 hectares of neighbouring Goedemoed brought the total land supplying the cellar to 107 hectares.

When Hennie arrived at Van Loveren the orchards contained oranges, guavas, pears and apples. The vineyards were planted with Muscadel, which he used to make sweet wine in the existing cellar, believed to be the oldest in the area. Hennie Retief was

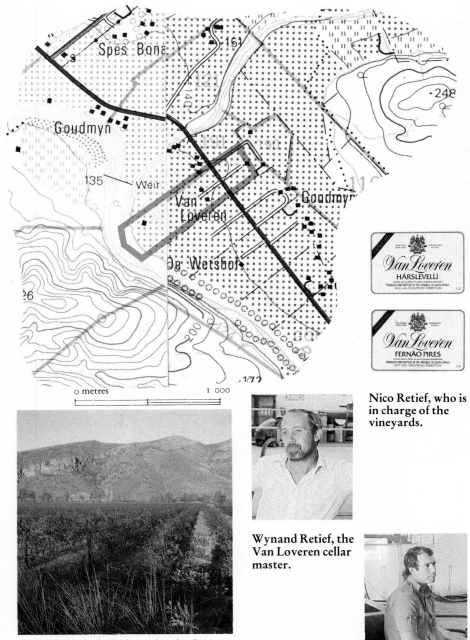

Nico Retief, who is in charge of the vineyards.

Wynand Retief, the Van Loveren cellar master.

The cleft in the hillside is a kimberlite deposit, a diamond pipe.

the first farmer in Robertson to rail wine in casks to Cape Town.

In 1942 the Breede River came down in flood, destroying the orange orchard. In the following years peaches, apricots and more Muscadel took the place of the pears, apples and guavas. Hennie continued to make sweet wine from Muscadel until his sons returned to take over the farming duties. Nico, the eldest son, finished his studies at Elsenburg and started working on the farm in 1961. Wynand finished at university in 1967. In 1982 Hennie died and his sons now farm Van Loveren as a partnership.

The vineyards around the cellar are planted in very fertile Kogmanskloof alluvial soils. Schoemanshoek and its neighbour have alluvial, sandy clay soils that are not as rich as those of Van Loveren. Goedemoed has the area's typical limestone outcrops. All three farms are part of irrigation schemes and the 200 mm of rainfall that falls in an average year is supplemented by drip irrigation.

In 1972 the cellar was extended to assist the production of dry wines, and cold fermentation equipment was installed. Wynand was given the responsibility of making the wines.

Nico and Wynand have replaced a number of orchards and old vineyards with new grape varieties chosen for their suitability to make fragrant dry wines. They've planted many new varieties, including Sauvignon blanc and Weisser Riesling. "We've made wines from Colombard, Muscat de Frontignan, Hárslevelü and Fernão Pires," says Wynand, "and these flowery, fresh wines are our style."

Nico and Wynand work the farms as a team, with Wynand moving into the cellar for three months of the year to make the wines. Their Hárslevelü and Fernão Pires were the first wines from these Hungarian and Portuguese varieties to reach the market in South Africa. They have a fresh, clean flavour, with a prominent bouquet.

Wine is for sale on the the Estate.

Weltevrede

As the sun sets on the Breede River vineyards, Colombard grapes begin the journey to the crusher.

When ideally ripe, red Muscadel grapes have a slightly wrinkled skin.

The Weltevrede Estate is situated roughly in the middle of the broad flat alluvial valley known as Bonnievale, some 25 kilometres south-east of Robertson. The expansive, comfortable house is surrounded by more than an acre of immaculate gardens and tall trees. And this in turn is encompassed on all four sides by row after row of flat vineyards. Throughout the broad expanse of vineyard lies a pattern of canals, all gravity-fed with precious water from the Brandvlei Dam on the Breede River. Bonnievale, on the border of the Robertson wine area, has an average rainfall of only 200 mm a year, making irrigation essential for the growth of any crop.

Bonnievale is one of the final wine-growing areas one encounters when travelling south-east from Worcester, and is one of the first to receive the cooling effects of the summer south-east wind, channelling up the valley of the Breede River on its journey from the southern ocean. The south-easter decreases evening temperatures, lowering the average temperatures and extending the length of the ripening season.

The valley's soil, rich with alluvial deposits from the Breede River and its Kogmanskloof tributary, provides a rich growth medium for the development of vigorous vines. This fertile

Lourens Jonker, pioneer in top quality white wines in the Breede River region.

base with high pH levels, together with the long, warm, ripening season, produces low pH grapes with adequate sugar, resulting in full-bodied wines.

Weltevrede has five basic soil types, extending from the coarse-grained red Karoo soils through pebbly loams to the fine-grained alluvial soils on the river bank. Most of the soils away from the river silt have a layer of clay beneath the top soil that is used to enhance the moisture-retention capability of the earth and provide the root system with wider horizons for ex-

Good quality juice is extracted with horizontal basket presses.

pansion. Preparation of a vineyard on Weltevrede starts with the thorough mixing of the top section of the clay with the top soil, which, in the case of the red Karoo soils, is up to two metres deep.

One of the most critical factors influencing quality on the Weltevrede Estate is the judicious use of irrigation to supplement the shortage in natural supply. Historically, the Weltevrede Estate has fed vineyards with flood irrigation using gravity drainage from the canal. Through this system sluice gates were opened to allow water to course through the vineyard, soaking into the soil around each vine. But during the '70s Lourens Jonker installed drip and micro-spray permanent systems in most of his vineyards, and is now able to give all his better

Colombard is used to produce high-acidity blending wines.

quality cultivars a controlled quantity of additional moisture when it is required.

With few exceptions farmland around Robertson and Bonnievale was of little value until well into the 1900s. Until the Brandvlei Dam was built, low rainfall restricted agriculture to the growing of patches of irrigated crops near rivers during winter, and to the grazing of sheep, goats and ostriches.

Weltevrede was originally part of a 2 555 hectare loan farm called Langverwacht, granted to Jacobus du Preez and Hendrik Wessels in 1831. In 1911 Hendrik van Zijl sold 380 hectares of this land to Nicklaas Jonker, Lourens's grandfather. Klaas Jonker, as he was known, divided his farm equally between his four sons, so each son received 95 hectares. Japie, Lourens's father, inherited the part that contained the family homestead, Weltevrede, in 1933.

This was during the time of the Great Depression and, for many years after, like farmers throughout the world, the Jonker brothers had to battle to make a meagre living. Vines from the Muscat family had been planted by their father, Klaas, and, though they made sweet wines, fortified with their own distilled spirit, there were many years when the wine had to be run off out of the cellar, into the ground. Few people had money to buy wine. Later, when the economy had recovered, Japie and his brother Herman, who had inherited the farm

Snow falls in winter on the Riviersonderend Mountains.

known as Muscadel next to Weltevrede, together built a cellar sited on the border of the two farms. As their business grew, the cellar was divided.

In the now separate cellars the two brothers each made sweet fortified wines: Muscadel, Hanepoot and Brown Sherry. During the 1960s, Lourens did agricultural courses at Stellenbosch and in California, and worked with his father on Weltevrede. In 1968 he decided that further expansion of the old cellar was impractical. It was pulled down and replaced with a cellar designed for the production of quality white wines, containing the most modern and efficient crushing and cooling equipment available.

In May 1969 Japie Jonker died, and Lourens took over Weltevrede. Shortly afterwards, in August 1969, Lourens bought the farm Muscadel from his uncle and incorporated it into the Weltevrede Estate. At this stage all the vineyards were on flat land, and the new era began with the

Weltevrede's modern cellar stands beside the irrigation canal.

incorporation in 1981 of a small adjoining farm, Riversedge, with south-facing slopes above the sweep of the river.

The two major principles that guide Weltevrede's pursuit of high quality wines are the care with which supplementary water is added to vineyards and the restrictions on the quantity of grapes produced by each vine. However, Lourens believes that small crops are not an essential part of quality production on high-potential soils, like those found on Weltevrede. He has demonstrated that when vines are restrained from producing their greatest potential quantity, there is a distinct improvement in quality, and he is convinced that the restriction on capability is more important than an ultra-small crop.

The soils and climate of the Bonnievale area have proved so well suited to the growing and ripening of Muscadel vines that the production of fortified dessert wines is a long-established tradition in the cellars of the region. The production of fortified wines has dropped to about five per cent of Weltevrede's production.

In 1969, when Lourens amalgamated the two adjoining Jonker properties, he found that the majority of the vines were Chenin blanc and Palomino, used for the production of sherry. Aware that the future of his enterprise lay in the successful production of dry and semi-sweet table wines, he planted Colombard and fermented the juice of both these and the Chenin blanc grapes to make the first flag-bearing wines of the Estate. To provide premium quality for the traditional dessert wine market, he bottled small quantities of red Muscadel from the 1976 vintage, and became the first wine Estate to market this type of wine.

During the late '70s Lourens began a major planting programme designed to provide classic quality grapes from noble varieties. The first of these to be established in the vineyards was Weisser Riesling, followed by Kerner, Sauvignon blanc, Gewürztraminer and Chardonnay. These varieties will gradually take over the carrying of the Estate banner and provide both pure cultivar wines and dry and semi-sweet blends. To further vary the blending spectrum, Lourens has blocks of Cape Riesling, Muscat de Frontignan and Therona Riesling, in addition to Colombard and Chenin blanc.

Cellar work and the intricacies of blending have long played a major part in the Estate's success. In 1979 Lourens introduced Privé, a Colombard–Chenin Blanc blend that was one of the few dry blended whites to be awarded Superior classification during that decade. From Colombard and Muscat de Frontignan wines, made during the following harvest, he originated Weltheimer, a semi-sweet blend. In 1979 he put some Privé into small casks of French oak to add a wood-matured dry white to the range, called Privé du Bois.

The 1981 vintage, after flooded vineyards, rain during harvesting time and persistent cloudy skies, brought great problems to the Weltevrede cellar, but contributed some interesting wines. The abnormal humidity and the cool conditions provided an unusual quantity of grapes infected with noble rot. Lourens and Inus Muller, his cellar master at the time, made Noble Late Harvest and Special Late Harvest wines with deep and complex flavours, using berries shrivelled and dehydrated by the infection.

Weisser Riesling produces a light crop of tightly-packed bunches on Weltevrede. Lourens believes that this variety should be picked in the cool of the morning and the grapes chilled before crushing. After the skins are broken, they are allowed to remain mixed with the chilled juice for at least one hour before they are pressed, to obtain all the available juice. This liquid is then clarified through a centrifuge before inoculation with a multi-culture of wet yeast and, like all of the Weltevrede white wines, fermented at relatively cold temperatures (11-12 °C). Musts that will be used for the production of semi-sweet wines are chosen through regular tasting and evaluation while fermentation is under way. Fermentation is arrested at the chosen degree of sweetness by the removal of the yeast from the liquid in the centrifuge, after which the new semi-sweet wine is filtered. The other wines are fermented dry, then cleaned and bottled without delay. The one exception is the wine destined for wood maturation, which will have several months of oak contact as an intermediary stage.

Wine is for sale on the Estate.

Weltevrede's growing conditions ripen Muscadel grapes to high sugar levels and produce fine dessert wines.

Wonderfontein

The Breede River, flowing strongly during winter, dwindles to a mere trickle in summer. This forced the eighteenth and nineteenth century farmers, cultivating the fertile soil along its banks, to graze cattle and sheep to subsidise their unreliable return from the soil. There were therefore few vineyards north of the junction with the Kogmanskloof River, as vines planted in the deeply drained soils beside the Breede needed additional water in summer, and the few irrigation furrows usually ran dry just when the vines needed assistance most. Between 1880 and 1914, while ostrich feathers were fashionable in Europe, these birds provided a source of income that was much less dependent on the water flow in the river. When the world went to war in 1914 the ostrich trade died out and farmers had to look for other revenue.

J. P. Marais grew vines and made a small income from raisins on his family's farm, Wonderfontein, just outside the village of Robertson. Like other farmers in the valley, he suffered under the handicap of long, dry summers. To find a more regular source of water, he mounted a horse and rode upstream along the course of the river, searching for a suitable place for a dam. He rode almost as far as Worcester before he found a place where the course of the river could be blocked for a major storage dam. He then publicised his discovery and persuaded the government to finance the construction of the Brandvlei Dam. Today the majority of South Africa's wine is produced on the alluvial plains of the Breede River Valley, watered by canals fed from the Brandvlei Dam.

Though most of his income was provided by the cultivation of sultana grapes to make raisins, J. P. Marais built the first cellar on the property and made sweet wine, dry wine and distilled brandy. He was the recipient of the first agricultural distiller's licence in South Africa. In addition to Wonderfontein, he owned the farm Klipdrift, further south along the Breede River. When J. P. died in 1922 his son Kosie inherited most of Klipdrift and continued to sell brandy. A second son, Ernst, inherited Wonderfontein and the family cellar built in 1913. In 1927, Ernst died in a flu epidemic, and the third son, Eksteen, who had inherited part of Klipdrift, took over Wonderfontein.

Eksteen Marais was a dynamic character. During the Depression years, surrounded by farmers unable to sell their wine, he bought grapes from an ever widening circle of suppliers, made the juice into wine, and established a national mail-order market for his products. After the first year, J. P.'s cellar proved too small, and Eksteen began building a second cellar that was to be extended each year for several decades. The process of adding two or three tanks each year was only halted by the rise of the grape-growers' co-operatives after the Second World War. Most of Eksteen's suppliers joined co-operatives and he was unable to buy their grapes. By this time he had over two million litres of storage space in the form of concrete tanks, and a thriving market for the wine they contained. He was obliged to buy wine from the co-operatives to blend with the produce of the Klipdrift and Wonderfontein vineyards to be able to cope with the demand.

Most of the wine he made in the

The Wonderfontein vineyards are planted in loamy, alluvial soils on the river flats between the Breede River and Robertson.

constantly extended farm cellar was of the sweet, full, fortified type, as well as port and sherry. A smaller quantity of dry wine was made for more immediate consumption. Eksteen Marais was an entrepreneur and businessman. Faced with the prospect of selling wine where others had failed, Eksteen travelled to the Transvaal where he placed advertisements in daily newspapers. He encouraged readers to order his individually made wines from Wonderfontein. The majority of the orders were for sweet wine supplied in 22-litre wooden casks, railed to the customer's address. Wine was also supplied in 200-litre casks, mostly to hotels and restaurants. Eksteen imported used casks from the United States. They were shipped as staves and Eksteen employed a team of coopers to repair any damaged staves and reassemble casks to keep the orders rolling.

When the demand for sweet wines began to decline in the 1960s, Eksteen increased his production of dry wine and began to supply it in bottles. While Eksteen's son Paul was studying wine making at Stellenbosch University, Eksteen became very ill. Paul returned to the farm to run the cellar and the extensive liquor marketing business that had developed. On his father's death in 1971 Paul was left with a substantial business and a rapidly changing market. While sweet wine sales have declined to provide no more than 5 per cent of Wonderfontein's cellar income, sales of dry wine have increased to balance this decline.

But Paul has been faced with another problem of potentially major proportions. His dry wines have been made from the produce of vineyards planted by his father, and though the demand for these inexpensive '*ordinaire*' wines has kept the business running, he expects sales for these wines to be affected by the growing demand for high quality wines. He has therefore started to change the course of the operation developed by his father, and has planted vineyards of Weisser Riesling to bolster the present production of Chenin Blanc, Clairette Blanche and Palomino dry wines. Muscadel, produced in the style made famous by his father, will also be bottled as Estate wine.

The Wonderfontein Estate contains two sets of vineyards growing in different situations and producing different styles of wine. The 47 hectares of the Wonderfontein farm itself have vineyards on sandy loam alluvial soils, producing abundant crops of healthy grapes with little effort. The 30 hectares of vines on the Klipdrift farm have sandy soil with major deficiencies. This farm requires heavy fertilisation to produce good crops. Paul expects his best dry wines, as well as his top sweet wines, to be produced on Wonderfontein's richest soil, furthest from the river.

The sweet wines are made in the traditional style. The grapes for Paul's dry white wines are picked at an early stage of ripeness. After crushing, the juice is drained off the grape skins, inoculated with yeast and fermented at 15 to 16 °C until dry. A tank of unfermented grape juice is retained to sweeten Wonderfontein's semi-sweet wines.

These Estate wines are bottled and sold on the farm, together with a full range of other liquors from Wonderfontein's fully licensed premises.

Wonderfontein's traditional market for Muscadel wines, supplied in 5-litre containers, continues, but quantities have decreased.

In addition to the recently available Estate wines, Wonderfontein continues to supply the market for less expensive wines.

Wonderfontein

Zandvliet

Zandvliet's best Shiraz vineyard shows luxuriant growth in spring as the new grapes begin to ripen.

The De Wet family have grown vines and bred horses on Zandvliet since 1860.

These richly coloured, tightly packed Shiraz grapes reflect the quality of the Zandvliet soil.

Exciting red wine is being made in Ashton, though only ten years ago few South African wine farmers would have considered this even remotely possible.

Ashton is in the Robertson area, receiving as little as 200 mm of rain per year. It has long been known for the production of brilliant racehorses, of high-quality dessert wines and of succulent peaches for the canning industry, but the idea of good, dry table wines coming from Ashton is very new.

The making of quality white table wines was the first major achievement of the Robertson area's wine cellars. These began to appear around 1965, but did not receive full consumer acceptance until the mid-1970s. White wine cultivars, such as Steen, had long been grown in the Robertson and Worcester areas, but high summer temperatures had resulted in instability and early oxidation in the early dry white wines. But the improvement observed in the white wines coming from the Stellenbosch, Paarl and Tulbagh cellars equipped with cold fermentation facilities meant that it was simply a matter of time before refrigeration cellars were tried 'north of the mountains', as they say in the Cape.

It is now generally accepted that highly trained cellar masters in the inland areas are able to produce quality white wines from cultivars such as Steen and Colombard. These cultivars had been planted for purposes other than table wine production, such as sweet wine and rebate wine for brandy.

But red wine was another story. It had long been believed that red wine could be made successfully only near the sea. Another factor that hindered experimentation with red wines in these areas was the lack of quality red cultivars already in production in the irrigated areas, which could be used for experiments. But as in all new ventures, somebody had to break the ground; there had to be a pioneer.

The De Wets of Zandvliet, just a few kilometres south of Ashton, are born pioneers. Their forefathers came to the area and bred horses, with good results, and further success came with the production of sweet wines and finally of dry wines.

Zandvliet is a father and son story. Paul, the Estate's vintner, took over the cellar in 1971, and began to implement ideas conceived by his father, known as Paulie. Paulie had long studied the Estate's Shiraz vineyard, originally planted by his own father before the First World War. Shiraz was grown on Zandvliet through that period for the production of port. Paulie noted that the skins of Shiraz grapes turned almost black in summer and, as the grape juice had a high sugar content when ripe, there seemed little reason why it should not produce a richly flavoured, deep-coloured, dry red table wine. Paul, his son, enthusiastically set about the project of making a quality red wine. He selected Shiraz stock from the Meerendal Estate in Durbanville, long famous for its production of wine from this cultivar, in preference to using material from the Zandvliet Shiraz vineyard.

From this new stock Paul developed his own nursery of Shiraz planting material and planted Shiraz vineyards on the rich alluvial soils along the Kogmanskloof River and the limy red Karoo soils on the farm's higher land. "We get quite a lot of moisture from the sea," says Paul, "far more than most people imagine. We may not get it in the form of rain, but throughout summer a cool breeze blows from the south-east. Taking a direct line in this direction, we're less than 100 kilometres from the sea. I'm certain that this cool breeze and the moisture it brings are reasons why Shiraz produces well in this area. And if there are people who think that even 100 kilometres is too far from the sea, let them look at a map of France. There are quite a few districts in Europe that produce respectable red table wine and yet are situated further from the sea than we are. Our wine is not a Stellenbosch red, and I don't expect it to be judged as such. We produce a Zandvliet red and I'm prepared to stake the future of this Estate's cellar on its quality."

The first De Wets came to the Robertson wine area and settled around Ashton in the early 1800s. Until the Brandvlei Dam was built on the Breede River in the early part of this century they obtained a little water for crop production from springs high in the mountains, but they made most of their living from dry land farming. Later when the Dam was opened and a canal was built through to Ashton, vines were planted, fruit orchards were established and fodder crops, such as lucerne, became major sources of revenue.

There have always been horses on Zandvliet. Paul de Wet's great-grandfather raised and trained thoroughbreds, hackneys and saddle horses. Paul's grandfather, also Paul J., continued the tradition after he bought the farm from his father in 1902. And finally, Paul's father, Paulie, who took over the farm in 1947, developed the farm into a horse stud of international renown. When Paulie took over there were varied types of horses on the farm. Outstanding racehorses had been produced on the property and Paulie believed that the farm's Karoo soils, rich in lime, gave a growing horse that little extra that makes a good horse into a champion. He culled everything but the very best from the stock, imported some of Europe's finest blood stock and soon became one of the most famous racehorse breeders in South Africa.

In 1947 vineyards were the farm's most important income earner. In Zandvliet's old cellar Paulie continued to make sweet wines and wine for re-

Snow falls on the mountains adjacent to most of the inland vineyard area of the Cape.

Looking across the rich alluvial soils in the Kogmanskloof valley to Zandvliet in the centre of the picture.

bate brandy. Though horse breeding was his prime interest, he retained the better vineyards. In 1965 he built a modern wine cellar containing cold fermentation equipment, and pioneered the making of dry white table wines in the Robertson area.

When Paul, his son, finished at Stellenbosch University they began to run Zandvliet as a two-man operation. Paulie continued to give most of his attention to horse-breeding and Paul took charge of the vineyards and cellar. In 1978 they were joined by Paulie's second son, Dan, who assumed responsibility for the running of the horse stud.

Sweet dessert wines began to play a smaller part in the Zandvliet cellar's production than they had done in the past. Some of these vineyards were old, so were uprooted and planted with quality dry table wine varieties. New vineyards were established on what was previously either orchard or grazing-land. The whole farming operation with regard to techniques and equipment was studied and modified. Paul decided that drip and micro-spray irrigation were far more effective in his naturally deep-draining vineyards than the flood irrigation that had been used by both his grandfather and his father, so permanent irrigation is now installed in all vineyards. Of Zandvliet's 225 hectares under irrigation, almost 100 hectares are vineyard, the rest being planted with orchard and fodder crops.

The first Shiraz vineyard was planted in the early '70s and Paul made experimental wines with the first two small crops. He had never made dry red wine in commercial quantities before, and there was no one to teach him how to produce quality with individuality under his unique conditions. In 1975 the first full crop was harvested and Paul made his first tank of Shiraz wine. The wine was critically examined and found to be wanting. As the experts had predicted, Zandvliet Shiraz, made from grapes grown on fertile soils, under irrigation and ripened in Robertson's hot summer, was too light to be a serious wine. The wine was sent to Stellenbosch, bottled and stored. At that stage, nobody knew what would become of it. Paul was determined to allow Zandvliet every chance to show that great red wines could be made on the farm, and believing that the '75 grapes were picked too early, left the '76 grapes to ripen further.

While waiting for the Shiraz to reach higher sugar, Paul found that his Muscadel, the source of the major part of the farm's income from the vineyards, was ripe and ready to pick. This is a problem that now arises almost every year. Priorities have to be decided. In 1976 he decided to take the Muscadel grapes off the vines first. When he reached the Shiraz, he found samples of grapes measuring as high as 26 ° Balling, and thought that he had made another costly mistake, to provide the critics of Robertson red wines with further arguments. The Shiraz wine of '76 was dark and heavy, but surprisingly few tannins were evident in the taste. The wine was also matured and bottled in Stellenbosch. In 1979 the '76 Shiraz was certified Superior, thus putting Zandvliet on the map as a red wine producer. Less than a year later the '75 wine was also certified Superior and the public had the chance to compare the light and heavy Zandvliet styles. The general preference has since been for a compromise. The more recent vintages have had deep colour, moderately high alcohol, but few tannic, harsh tastes, even at three years.

Paul is still evolving his technique to get the most from his Shiraz crop. "The wine that we produce here is basically different from the Shiraz anywhere else, because we have to ferment longer on skins to get the colour we require. Even by fermenting down to 5 ° Balling at 25 °C, the wine is soft and not astringent," he says.

"We aim for an average Balling content of about 23½ °. Now that the crop is much larger, we have to start picking earlier and finish later, but we've always been able to get the grapes in at the right sugar."

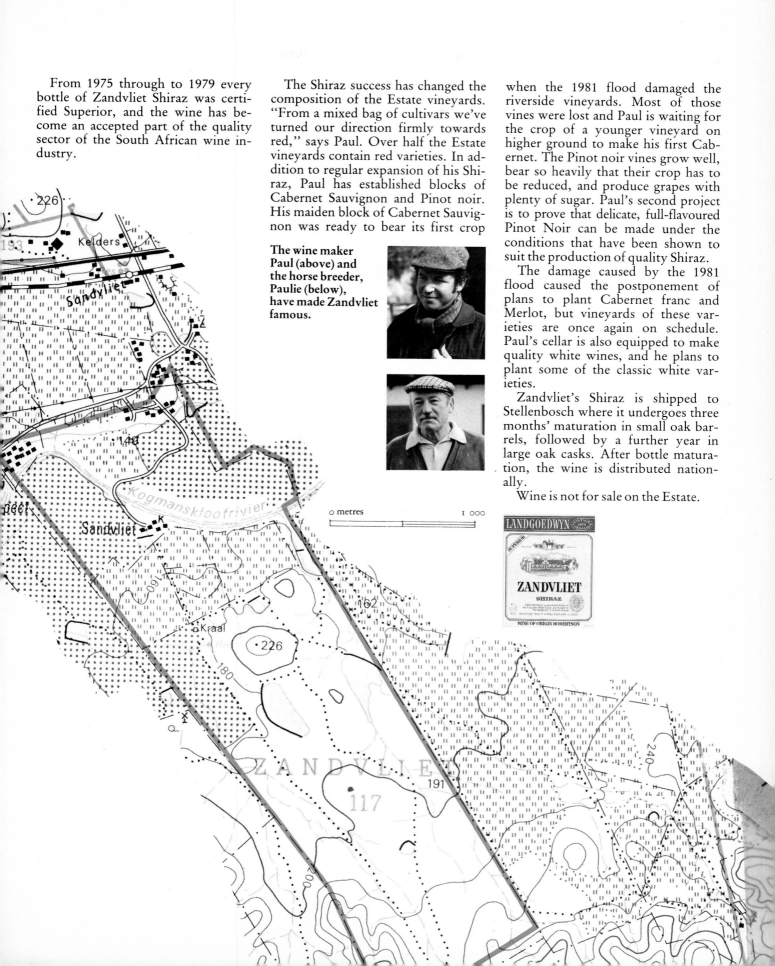

From 1975 through to 1979 every bottle of Zandvliet Shiraz was certified Superior, and the wine has become an accepted part of the quality sector of the South African wine industry.

The Shiraz success has changed the composition of the Estate vineyards. "From a mixed bag of cultivars we've turned our direction firmly towards red," says Paul. Over half the Estate vineyards contain red varieties. In addition to regular expansion of his Shiraz, Paul has established blocks of Cabernet Sauvignon and Pinot noir. His maiden block of Cabernet Sauvignon was ready to bear its first crop when the 1981 flood damaged the riverside vineyards. Most of those vines were lost and Paul is waiting for the crop of a younger vineyard on higher ground to make his first Cabernet. The Pinot noir vines grow well, bear so heavily that their crop has to be reduced, and produce grapes with plenty of sugar. Paul's second project is to prove that delicate, full-flavoured Pinot Noir can be made under the conditions that have been shown to suit the production of quality Shiraz.

The damage caused by the 1981 flood caused the postponement of plans to plant Cabernet franc and Merlot, but vineyards of these varieties are once again on schedule. Paul's cellar is also equipped to make quality white wines, and he plans to plant some of the classic white varieties.

Zandvliet's Shiraz is shipped to Stellenbosch where it undergoes three months' maturation in small oak barrels, followed by a further year in large oak casks. After bottle maturation, the wine is distributed nationally.

Wine is not for sale on the Estate.

The wine maker Paul (above) and the horse breeder, Paulie (below), have made Zandvliet famous.

120　*Stellenbosch*

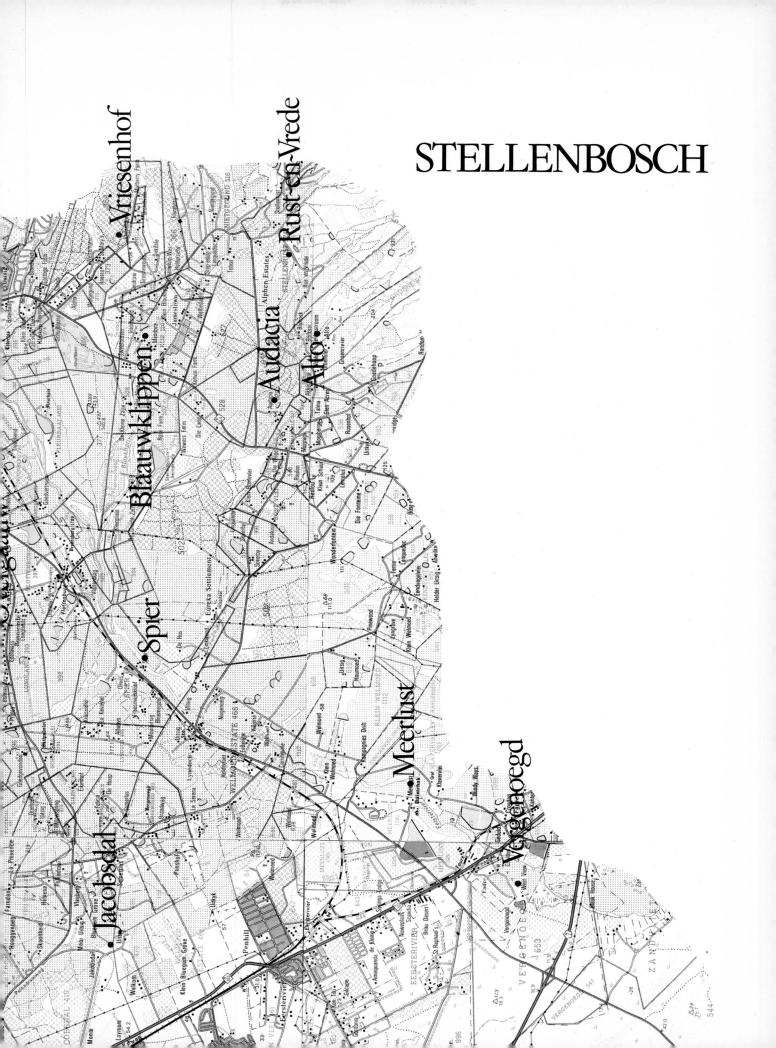

Alto

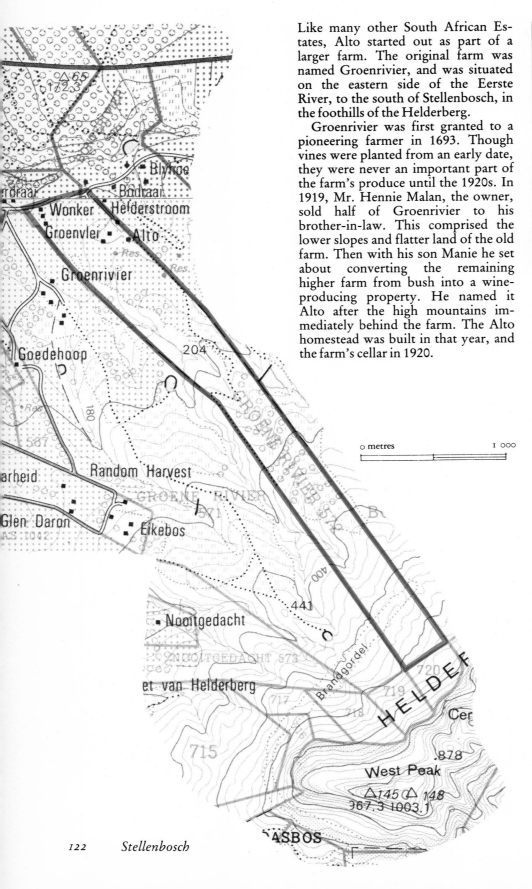

Piet du Toit, Alto's vintner. His son Gerhard (Hempies) is the Estate cellar master.

Like many other South African Estates, Alto started out as part of a larger farm. The original farm was named Groenrivier, and was situated on the eastern side of the Eerste River, to the south of Stellenbosch, in the foothills of the Helderberg.

Groenrivier was first granted to a pioneering farmer in 1693. Though vines were planted from an early date, they were never an important part of the farm's produce until the 1920s. In 1919, Mr. Hennie Malan, the owner, sold half of Groenrivier to his brother-in-law. This comprised the lower slopes and flatter land of the old farm. Then with his son Manie he set about converting the remaining higher farm from bush into a wine-producing property. He named it Alto after the high mountains immediately behind the farm. The Alto homestead was built in that year, and the farm's cellar in 1920.

From that first year the Malan family set out to make only red wines, and during 1920 Hennie Malan planted Cabernet Sauvignon, Shiraz, Pinot noir, Gamay and Cinsaut. Gamay did not grow well, and these vines were soon replaced. Pinot noir grew well, but its early-ripening properties counted against it on Alto. It was past its best when the Cabernet, Shiraz and Cinsaut vines were ripe and able to be crushed together to make a naturally blended wine.

Hennie's son Manie had been among the very first students to study agriculture at Stellenbosch. "Manie was a man of exceptional ability," says Piet du Toit, vintner and part-owner of Alto. "Without the soil tests and other agricultural aids of today, Manie had wine farming taped a long time ago. When he returned from University he chose the part of Groenrivier that was best suited to red wine farming. Then he chose to plant the noble varieties.

"He didn't have a lot of money. He couldn't afford to keep wines for long maturation periods in the expensive French oak vats in his cellar. With white wines, there wouldn't have been a problem. In those times a white wine made in South Africa was considered past its peak after two years. But wine made from Cabernet Sauvignon, for example, needed at least four years' maturation before it was ready for drinking. That was the

122 Stellenbosch

4 000-litre vats are used to mature most of Alto's red wines. These have been fitted with Manie Malan's 1920 two-inch brass taps.

reason for the birth of Alto Rouge."

Manie decided that he would blend Cabernet Sauvignon with wines that matured much earlier, such as Cinsaut and Shiraz, to get a smooth, round wine that was fit for the table after two years' maturation.

The fact that each of the varieties ripened at much the same time allowed them to be crushed and fermented together to make a natural blend. The blended wine they made was of such quality that it won first prize in the 'dry red wine, Burgundy-type' class at the Cape Wine Show for six consecutive years from 1924 to 1929. In 1924 Manie sent a sample of the farm's wine to a leading London wine merchant, Burgoyne's, who were so impressed with the wine that they contracted to buy a portion of each year's crop for the next five years. Alto supplied Burgoyne's with the wine we know as Alto Rouge from 1924 to 1962, when Die Bergkelder became responsible for marketing the farm's wine.

In 1934 Burgoyne's began to buy white wines from Twee Jongegezellen and Theuniskraal in Tulbagh, and they appointed Manie as their representative in South Africa. He had to judge and choose the quantity of Tulbagh white wines that Burgoyne's were to import. Alto Rouge was first bottled for South African sale in 1933, nine years after the English had their first taste of the farm's wine.

Piet du Toit came to Alto as manager and wine maker in 1959. He was new to the district and knew little of the traditions on Alto. He called on the famous Professor C. J. Theron, who was at that time head of the faculties of viticulture and oenology at Stellenbosch University. Together they studied the farm from every angle. Alto is quite close to False Bay, and it has been a long-held belief that red wine should be grown close to the sea, to benefit from the moisture-laden sea breezes.

Alto is situated on the slopes of the Helderberg, and has the granite soils many consider necessary for the growth of high-quality red wines in South Africa. The Alto Estate slopes to the north-west and receives the warm afternoon sunshine needed to get the best from red cultivars. Together, Professor Theron and Piet du Toit concluded that Manie Malan had not made a mistake in his development of Alto as a red wine farm. When they surveyed the farm, 60 per cent of the vines were Cinsaut, 20 per cent Shiraz, and 20 per cent Cabernet Sauvignon. These grapes were harvested together, if the season allowed, and the juice was blended as the grapes went through the crusher.

In those early days, Alto Rouge was a Cinsaut-based blend, light-bodied and early-maturing. In 1970

Alto Rouge being weighed before shipment to London. The wine was exported to Burgoyne's from 1924 to 1962.

Alto's long-matured Cabernet Sauvignon is being made in a slightly lighter style than earlier releases.

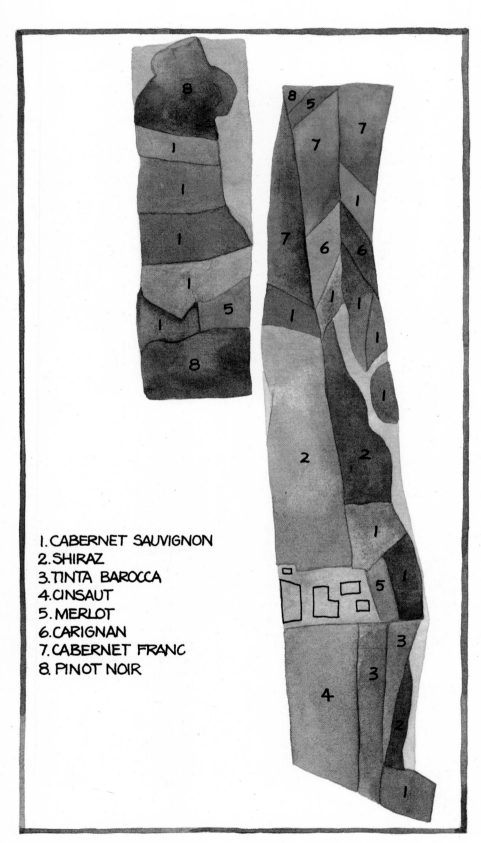

1. CABERNET SAUVIGNON
2. SHIRAZ
3. TINTA BAROCCA
4. CINSAUT
5. MERLOT
6. CARIGNAN
7. CABERNET FRANC
8. PINOT NOIR

Piet began to plant Shiraz vines intending gradually to increase the proportion of this variety in the blend.

In 1965 some of the Cabernet Sauvignon grapes were fermented separately. They had been allowed to reach late-ripeness, thus producing a very sweet juice that was fermented in open concrete tanks, with the skins in the must until all the sugar was fermented out. Heavy Cabernets in this style were made each year with two exceptions until 1971, when a shortage of Cabernet Sauvignon in the vineyards caused its suspension. The wine was aged in large oak barrels for three years and bottle-matured for a further three years before marketing.

The separation of some Cabernet and the introduction of more Shiraz meant that in the 1960s Alto Rouge became a Shiraz-based blend, and remains so at the time of writing. During this period, Manie Malan's old Cabernet Sauvignon vineyards became due for replacement and new vineyards of the same variety were planted in 1971 and 1972. The subsequent period of vine growth before the maiden crop of grapes was responsible for the suspension of Alto Cabernet Sauvignon after 1971. The new vines produced their first crop in 1975, and the resumption of the maturation of Alto Cabernet Sauvignon was recommended the following year. However, the wet year of 1977 followed and Alto was unable to ripen Cabernet Sauvignon sufficiently and once again the wine was not released.

Another replanting programme was begun in the late 1970s that may eventually again change the composition of the Alto Rouge blend. Large blocks of Merlot, Cabernet franc, Carignan and Pinot noir have been planted in addition to more Cabernet Sauvignon. Over 40 hectares of the Estate's 100 hectares of vineyard have been replanted in this programme.

Alto is a very long farm, 2,5 kilometres in extent and never more than a few hundred metres wide, with the most elevated vineyards 300 metres higher than the lowest ones. An unusual feature is the extended rows of Shiraz, more than 1,8 km long in the centre section of the farm. The vines were planted in this manner to facilitate mechanical cultivation of the ground between the rows.

The only modifications made by Piet du Toit to Manie Malan's cellar during the period 1960-80 involved the replacement of old maturation vats and the addition of some extra storage space. However, when the small 1975 crop of Cabernet Sauvignon was ready for harvesting, Piet changed the fermentation technique used for this variety. To produce a lighter style of wine and reduce the heavy tannins evident in the previous vintages, he removed the skins from the fermenting must with a little less than half the sugar remaining. This remained the practice until 1981, when a new pressing cellar with stainless steel closed fermentation tanks was built at Alto. Once again the Cabernet Sauvignon fermentation technique has been modified to retain the lighter style, but with greater flavour extraction. Today the temperature of the fermenting must is maintained at about 25 °C, with the skins present in the must until all or almost all of the sugar has been converted. As the farm's red wine production has increased recently, the Cabernet Sauvignon is only aged for one year in the Estate's maturation cellar until the next year's crop comes in, and is then moved to the Bergkelder, where it is placed again in oak.

In 1979 Hempies du Toit, Piet's son, returned to the farm to take charge of the cellar. The arrival of Hempies, a product of Stellenbosch University's oenological and viticultural course, has coincided with the introduction of the Merlot and Cabernet franc blending varieties, and Pinot noir. Planted in the rarefied atmosphere at the top of the farm, Pinot noir ripens very easily without losing acidity. With more than 40 000 vines planted high on the hillside, Alto should be able to produce significant quantities of this delicate, fragrant red wine for blending and possibly as a cultivar wine as well.

Another interesting newcomer in the Alto cellar is Carignan, the vol-

Looking east from Alto's sloping vineyards on the Helderberg toward Stellenbosch Mountain.

ume-producing red variety, popular with French and American vine growers. Hempies reduces the abundant crop by more than half before it begins to ripen, allowing the vine to direct the energies of its vigorous root system into concentrating the sugar and flavour compounds of the remaining grapes. With this technique, this variety produces grapes with high sugar acid content. The young wine is ruby red and matures much faster than the more noble varieties.

We await the produce of the Alto Merlot and Cabernet franc vineyards in whatever form the Du Toit family and Distillers Corporation, their partners at Alto, choose to provide.

Wine is not for sale on the Estate as Alto's wines are matured, bottled and distributed nationally by a Stellenbosch wholesaler.

Audacia

Sloping to the west, Audacia's prime vineyards are pruned to restrict production and achieve high quality.

In 1930 romance changed the fortunes of one of Stellenbosch's Estates and created another. Kosie Louw, eldest son of Neethlingshof's owner, wished to marry, and was unwilling to wait for his father's blessing. He left Neethlingshof and his inheritance to buy a small piece of the historic Annandale farm on the Stellenbosch–Strand road. This 16-hectare farm was a mixed vineyard-orchard. He then leased an extra 112 hectares of adjoining land, on the higher Helderberg slopes, from the Stellenbosch municipality to amalgamate with it. As this municipal land was bush, he had to clear the trees before he could use it even for grazing land. The 16-hectare farm, together with the leased grazing property, did not impress Kosie's friends as being a particularly exciting commercial proposition, especially as he had also forfeited his potential future inheritance of at least a part of Neethlingshof. One of Kosie's friends described the venture as nothing less than audacious, and from this Kosie derived the word Audacia as a fitting name for his fledgling enterprise.

In 1937 he was able to buy the municipal land and began to plant vineyards on it. The soil on this higher ground was a mixture of sand and loam, providing very good drainage for vine roots. The soil on the smaller portion of Audacia was deep, rich alluvial. The Bonte River formed its southern boundary, and over the years had brought a great deal of rich, dark soil down from the mountain range that lies to the east.

When Kosie first came to Audacia he found Sémillon, Cinsaut and Steen vines planted between apple and pear trees. Later, when the orchard had been taken out, the area was planted with Steen, the vine most likely to cope with the water in the low-lying alluvial soil in winter, and the lack of water in the same hard-packed soil in the summer. The extended vineyards of Audacia were planted with more Steen and Cinsaut, and a large area with Cabernet Sauvignon. His decision to plant Cabernet Sauvignon in quantity in 1935 was out of step with the rest of the industry, for this cultivar was not planted on a large scale by most South African wine farmers until the early 1970s.

The cultivar that proved most popular with Kosie Louw was Cinsaut, for on Audacia Cinsaut changed its character. Long known as the bulk-producing red cultivar of the South African wine industry, on Audacia's sandy soils Cinsaut produced a relatively small crop of high-quality red wine. Kosie was so impressed by the wine Cinsaut made on his farm that in 1957 his wines made their first appearance at the annual Cape Wine Show. He entered seven classes and won four first prizes, a second and a

Kosie Louw, maker of medium bodied red wines from vineyards on the lower slopes of the Helderberg.

126 Stellenbosch

third. One trophy for the best red wine of the Show was won by a Cinsaut, and another was awarded for the most points in the red wine section. Several of the prizes were won in the Cinsaut category in different quantity classes. (At that time, some of today's top Estate farmers exhibiting red wines at the Cape Show were the Roux family of Verdun and the Faure family of Vergenoegd.)

Kosie had by this time planted extra vineyards of Steen and Clairette blanche and made dry table wines from all the cultivars on the farm. In that year (1957) of Kosie's most successful wine show, his son Koos joined him on the farm. Since then, Koos has worked with Kosie developing the extra land on the farm and extending the area under both red and white cultivars. Because of its obvious liking for the farm, Cinsaut has been given priority in the area of land under new vines. Of Audacia's red vineyards 70 per cent belong to the Cinsaut cultivar. The area under Cabernet Sauvignon has not been developed to the same extent, covering about 10 per cent of the farm's red wine ground. Koos has been responsible for planting both Pinotage and Tinta Barocca, which make up the extra 20 per cent of the red wine crop. Koos, having also planted more Steen and a small area of Riesling, has reached the stage where Audacia has approximately equal areas of red and white vineyards. Because Steen is just as prolific a producer on Audacia as it is elsewhere in South Africa, the percentage of the crop, white to red, is roughly 60 to 40. As planting has gone ahead on a scale much greater than the development of Audacia's cellar, Koos makes only red wines on the farm, the white cultivar crop being sold to a wine merchant at grape stage.

Audacia was launched as a wine Estate with the 1974 vintages of Cabernet Sauvignon, Pinotage and Cinsaut.

On the lowest slopes of the Helderberg, Audacia's Hutton soils are well suited to the production of full-bodied reds.

These wines were first seen by the public at the 1976 Stellenbosch Bottled Wine Show. They had been matured for several months in small wood and then bottle-matured, and were acclaimed at the Show.

In 1977, the Cape had an extraordinarily cool summer with low temperatures and an unusual amount of rain. On 10 December Koos first noticed signs of white powder on the almost fully formed bunches of grapes, particularly among the Pinotage vines. He realised that this was evidence of an infection of downy mildew, and he intensified his dusting programme to reduce the effect of this unaccustomed scourge. However, within two days all his vineyards were seriously infected. When harvest time arrived, he found that only his crop of Steen was worth harvesting. The farm's Pinotage was left on the vines and the Cabernet Sauvignon did not reach sufficient ripeness to make a significant wine. The blow to the Pinotage became permanent when the depressed vineyard was unable to produce a crop of grapes the following year, and the vineyard had to be taken out.

Audacia has largely recovered from the blows of 1977. The Pinotage has been replanted and new vineyards of Steen, Sauvignon blanc and Tinta Barocca have been planted.

Wine is for sale on the Estate.

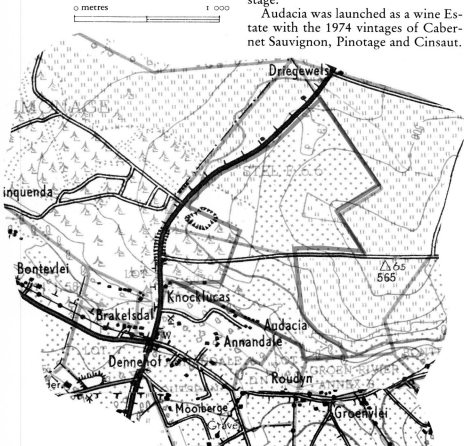

Audacia

Blaauwklippen

Blaauwklippen lies on gently sloping ground, at the foot of the Stellenboschberg.

Walter Finlayson has won many accolades for the wines he has produced in the Blaauwklippen cellar.

Densely forested, standing below the massive Stellenboschberg, Blaauwklippen first rang with the sound of iron axe on wood in 1682. In that year Gerrit Visser, a potter from Omnen in Germany, brought his wife and children over sand dunes and through dense brush and scrub from the Table Bay settlement to the uncleared farm that he named Blaauwklippen. Visser had been in the Cape in the service of the Dutch East India Company for seven years when he was promised 43 hectares of choice farming land in the virgin valley of Stellenbosch by the new Governor, Simon van der Stel. The potter, who was known locally as Grof (rough), marked out three separate, almost adjoining, pieces of cultivable land and set up house within a short horse-ride of the dozen fellow pioneer settlers in Stellenbosch.

Like most of the other new settlers, Visser had never farmed before. In the clearings he had made on the still heavily-forested farm he planted vines and raised horses, cattle, sheep and pigs. Each year he gave a pot that he had made in his farm kiln to the Governor at the Stellenbosch festival, held each October to commemorate Van der Stel's birthday.

Grof Visser farmed Blaauwklippen for eight years. In 1690, he petitioned Van der Stel for an official grant of the property that would allow him to sell it.

On 9 November 1690 Visser sold Blaauwklippen to Guillaume Nel, a Huguenot refugee. Nel brought his wife, Jeanne de la Batte, and two children, Jean and Jeanne, born in France, to the farm. Guillaume (later called Willem by his Dutch and German neighbours) and Jeanne eventually had eleven children, three of whom went on to own important Estates in the Stellenbosch district.

Jeanne, the eldest daughter, and her husband, Barend Pieterz van Wesel, inherited Blaauwklippen, Adriaan bought Uiterwyk and By-den-Weg (of which today's Overgaauw Estate was then a part), and Anna married the owner of Uitkyk.

The Hoffman family was the nearest to a dynasty that Blaauwklippen has seen. With the exception of a short period in 1771, the Hoffmans owned Blaauwklippen for more than forty years. Johann Bernhard Hoffman, born in Stralsund, Germany, came to the Cape in the service of the Dutch East India Company in 1744. He was Acting-Landdrost of Stellenbosch from 1747 until 1752, when he

The elegant Blaauwklippen home was built by Dirk Hoffman in 1789 to house his twenty-two children.

Low-lying alluvial soils beside the Blaauwklip River had to be provided with underground drainage before Sauvignon blanc was planted.

was given his free burgher rights and he bought Libertas, his first farm. In addition, he bought Spier in 1754 but sold it three years after buying Blaauwklippen in 1762. Hoffman was a leading citizen in the Stellenbosch district. In addition to fathering eighteen children, being nominated Heemraad and appointed Captain of the Burgher Dragoons (the citizens' volunteer defence force), he co-operated with another German ex-corporal, Martin Melck, in the furnishing of the original NGK in Stellenbosch with precious objects and the elimination of Stellenbosch's island. (The Eerste River had several times threatened to wash the tiny town of Stellenbosch from its perch on an island between two branches of the stream. After the high-water mark receded in 1769, the two ex-corporals supervised the filling in of the lesser of the two branches of the river, and the threat of future flooding vanished.)

Dirk Wouter Hoffman, son of Johann Bernhard, owned Blaauwklippen from 1773 to 1802, and raised twenty-two children on the farm. To provide shelter for his children he rebuilt his father's house and another earlier dwelling, creating the two fine, thatched houses that are still standing today in perfect condition.

As the quality of agricultural equipment available in the Cape improved, particularly in the nineteenth and twentieth centuries, Blaauwklippen began to slip back from the forefront of Stellenbosch farms. Blaauwklippen was named for the profusion of stones in its soil, which had been little handicap until the drive for greater production caused other farms to be fully cleared and planted, and exposed the difficulty of intensive farming in the stony ground. The stony patches proved to be too great an obstacle for a succession of owners until as recently as 1971, when almost all the surface stones were removed.

One of the many owners, Cecil John Rhodes, owned Blaauwklippen for less than one day. He bought it at an auction in 1898 for £3 500 and sold it later the same day for £3 000. The rapid changes of ownership ceased in 1971 with the purchase of the farm by Graham Boonzaier and a new chapter in the Blaauwklippen story began.

Graham, an industrialist with interests in the Transvaal and the Orange Free State, was the first modern-day owner of the farm with purposeful ideas for the re-establishment of producing vineyards and sufficient capital to allow the ideas to be translated into reality.

The farm was originally bought to give Graham a base to develop his ambitions in the production of limited quantities of cheese of Stellenbosch origin but this project was soon replaced by the vineyard reconstitution programme.

The unpruned, untended vineyards of 1971 covered about three hectares. They surrounded the ancient farm buildings, backed on to a square beside the Blaauwklip River and adjoined about a dozen hectares of apple orchard and a similar area of forest. Graham set out to recover as much of the land within the original Blaauwklippen boundaries as possible and today the farm covers almost 300 hectares.

After almost three centuries of exploitation, the soils were in a poor state, eroded and unfertilised. The period 1972-8 saw extensive soil reconstitution and vineyard planting changed the visual and economic faces of the farm. The higher soils are mostly Hutton in type and gravel in texture. The lower soils, near the river, are alluvial and have been heavily leached over the years. Some of the vineyards on the farm are irrigated during dry periods and the lower soils require additional water more frequently. The soils have been limed and fertilised according to the needs shown by analysis. In addition, they have had green and animal manure supplementation, and the effect has been seen in the vigorous growth of the vines and the relative ease by which the grapes reach the required sugar levels.

The second chapter in the story of Blaauwklippen's renewal opens with Walter Finlayson's arrival on the farm in 1975. Walter had been a wine maker on his family's Montagne Estate during the period 1960-75, and joined Graham at Blaauwklippen when Montagne was sold to Gilbey's. He brought a thorough grounding in the making of quality wines under Cape conditions and an understanding of the need for the retention of the natural delicacy of red and white quality cultivars under the warm, dry conditions of Western Province farming. His wines at Montagne were categorised as gentle, light, stimulating and racy, in contrast to the prevailing Cape preference for weight and fullness. This sensitive hand can be seen in the fruitier, but similarly delicate, wines of Blaauwklippen.

After Walter's arrival in 1975, a modern temperature-regulated cellar was built, using the old farm cellar of 1812 as a white wine tank cellar. The cellar has since been extended to house the total annual harvest of all the farm's vineyards. Because of the youth of the vineyards on the property when the cellar was being completed early in 1976, Walter and Graham had to obtain grapes from neighbouring farms to be able to make wine to supplement the income of the farm. The first small quantity of Cabernet Sauvignon from the Estate's own vineyards was harvested in 1976, and succeeding years saw vineyards of other red and white varieties come into production.

On the lower slopes of the Stellenboschberg, the red variety vineyards provide grapes with good colour, depth and balance.

Today Blaauwklippen has 220 hectares of vineyards, each one planted in balanced soils. About sixty per cent of the vineyards are planted to red varieties: Cabernet Sauvignon, Cabernet franc, Merlot, Shiraz, Pinotage, Pinot noir, Zinfandel and Pontac. More Cabernet Sauvignon vines are planted than any other. Cultivar wines are made from Cabernet Sauvignon, Shiraz, Pinot noir and Zinfandel. Red Landau, a blended red wine, is made from Cabernet Sauvignon, Shiraz and Zinfandel. A second blended red, the traditional Bordeaux blend, is made from Cabernet Sauvignon, Cabernet franc and Merlot. The sweet juice of late-picked Zinfandel and Pontac grapes is fermented down to about 7° Balling and then fortified to make an Estate port. The Pinotage grapes are picked at low sugar content and their

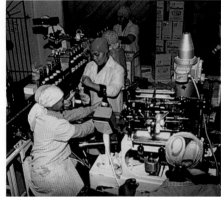

On Blaauwklippen, like most wine Estates, hand work often takes the place of machinery.

juice is separated from the skins immediately after crushing to make a lightly coloured white wine that is sold as Blanc de Noir.

Blaauwklippen's red varieties reach full ripeness without undue strain. The wines obtain sufficient colour and body without sacrifice of acidity. They have an attractive softness and fruit while young but are expected to age beneficially. As all of the red vineyards are still young, it is expected that the greatest of Blaauwklippen's wines will be reds of vintages still to come. Of recent vintages, the 1978,

Blaauwklippen's oldest vineyards date from 1972, and the quality of the wines has been improving as the vines gain maturity.

Old Cape carriages are on view and period furniture is provided for visitors to the tasting-room.

1980 and 1982 seasons produced outstanding wines, particularly Cabernet Sauvignon. In 1980 the farm's Zinfandel won for Walter Finlayson an award as the most innovative South African wine maker of the year. One can make a qualified assumption that Blaauwklippen, with its medium-bodied, fruity and complex red wines, may be best served by the Cape's warmer years.

Less than half the Estate's production is from the forty per cent of the vineyards that contain white varieties. These are planted to Chardonnay, Sauvignon blanc, Chenin blanc, Kerner, Weisser Riesling and Colombar. There are also small vineyards of Pinot gris, Muscat Ottonel and Bukettraube. It is expected that the more important dry white wines of the farm will be made from Chardonnay and Sauvignon blanc. To date, Weisser Riesling and Chenin blanc have produced wines of character, varying in sweetness from off-dry to Special Late Harvest. Small quantities of Noble Late Harvest have been made from blends of Pinot gris, Bukettraube, Sauvignon blanc and Kerner. Colombar is used to make low-alcohol, light, dry white wine for early consumption.

The period 1975-80 was influenced by the need to buy grapes to supplement the small production from the recently established vineyards, but today the farm vineyards supply most of the cellar needs. Blaauwklippen is far advanced towards improving on the image established during the bought-grapes period. It is now producing high-quality red and white wines, able to be matured for many years, carrying a stamp of consistency right across the range. The Cabernet Sauvignon wines, in particular, have a rich berry flavour with a breadth and depth that promises well for the future. The third chapter in the redevelopment of Blaauwklippen is just beginning.

Owing to the past policy of purchasing grapes, Blaauwklippen is not a registered Estate.

In addition to the limited production of noble wines, Blaauwklippen has become well-known for the production of farm foods. During the late '70s, Graham began to sell on the premises air-dried, wine-cured, salami-style sausage, pickles and other condiments and cheeses made from the Estate's dairy herd. Since 1978 these foods have been offered to the public in the form of an alfresco summer lunch, under the oaks.

In 1981 Graham began to produce commercial quantities of mushrooms from a factory built on Estate land next to the river, considered unsuitable for vines. Medallion Mushrooms have added a further dimension to the farm's already extensive list of attractions.

Wine is for sale on the Estate.

Blaauwklippen

Bonfoi

Bonfoi's Chenin blanc vineyards are on south-east facing slopes and level land in the western end of the Stellenbosch Kloof.

On Friday, 17 February 1970, the day picking was due to begin, the heavens broke over Bonfoi and a hailstorm destroyed virtually the entire crop. The little that was left to be picked was harvested in two days. The Van der Westhuizen family lost not only the 1970 crop, but tens of thousands of vines had to be replaced over the following seasons. And yet not one of Bonfoi's neighbouring farms was seriously affected.

Bonfoi means 'good faith', and on 18 February 1970, as Christoff van der Westhuizen surveyed his ravaged farm, there could not have been an attribute he needed more. In February 1933 Christoff's father had lost an entire crop to the devastating effect of the hailstone. Thirty-seven years later the son also learnt to respect the power of the forces of nature and the effects they can have on the destiny of any farmer.

The Estate is placed on the lower eastern slopes of the Bottelary Hills, approximately halfway between two other major wine Estates, Overgaauw and Uiterwyk. It was originally part of 'Aan-den-Weg', given to Jan van Rheede by Willem Adriaan van der Stel in 1699 and situated on the old Stellenbosch to Cape Town road.

Subdivided from 'Aan-den-Weg' in 1812, Bonfoi acquired this home in the same year.

Bonfoi gained its French title from the fact that at one time, soon after it was partitioned from the main farm, it was owned by two Frenchwomen. Later Bonfoi passed into the hands of the Joubert family, who owned large tracts of what are now extremely valuable winelands in the valley between the Bottelary Hills and the Helderberg Mountains. During the late 1920s a widow Joubert, who owned Bonfoi, married Paul la Grange, who had played rugby for South Africa in 1924, and the surname of the owner of the farm changed again. La Grange sold Bonfoi to Johannes van der Westhuizen, his foreman, for £12 500 in 1934.

Johannes van der Westhuizen was one of eight brothers and four sisters born to a Sandveld wheat farming and grazing family. He broke with family tradition and, like Petrus Bestbier, who became his neighbour at Goede Hoop, left the grazing country for the winelands during the Great Depression.

There were many who considered Johannes's decision to purchase such an expensive piece of land very foolish, and in 1935, his first year of operation, Johannes was able to repay only £700 of the money he had borrowed. But the bad years of wine farming in South Africa were coming to an end, and after extensive planting of vines and some good management, Bonfoi was able to pay off Johannes's debts and provide a healthy living for the Van der Westhuizen family. Christoff van der Westhuizen and his twin brother George worked for their father from 1956 until 1958, when they jointly purchased the farm. They then bought another farm on the other side of the Bottelary Hills and ran the two farms as a partnership, Christoff living on Bonfoi and George occupying the new farm. Later the partnership was dissolved and Christoff became sole owner and vintner of the Bonfoi Estate.

When Johannes van der Westhuizen bought Bonfoi in 1934, it was a

Christoff van der Westhuizen has owned Bonfoi since 1958.

fruit farm with a small area of vineyard. He had plans to make great wines, and within a short period of time had planted areas of Cabernet Sauvignon, Shiraz, Cinsaut, Steen, Riesling and Palomino vines. But Johannes soon found that demand for low-yielding varieties was negligible and he began to concentrate solely on Chenin blanc, Cinsaut and Palomino. One generation later, Christoff has reversed the process and replaced the Cinsaut and Palomino with Shiraz, Cape Riesling, Sauvignon blanc and Weisser Riesling.

Christoff started selling Chenin Blanc, a delicate and slightly sweet white wine, under the Estate label in 1976. "Chenin blanc is very well suited to this farm," he says. "The grapes ripen easily, with plenty of acid, and they make very fruity wines." He picks his Chenin blanc

Wines made from grapes grown on the slopes are blended with others made from grapes grown on level soils.

vines when the sugar content of the grapes is just over 20° Balling. Small quantities of grapes picked later at 22° Balling provide blending wine that adds body to match the sweetness and acidity of the wine.

The grapes are crushed and the skins are left soaking in the juice to add flavour for four hours. The juice is then drained off and fermented with a culture of yeast at between 14 °C and 16 °C. Only the free-run juice is used for the quality wine. As soon as the liquid has finished fermenting, it is rapidly cleaned and despatched to the Bergkelder in Stellenbosch, where it is bottled and bottle-matured before release to the market.

Eighty-five per cent of the Bonfoi vineyards contain white cultivars. Christoff has planted Sauvignon blanc and Cape Riesling to blend with his dry Chenin blanc for a new wine. "My farm seems to be best suited to white cultivars. They grow well and give me high quality juice to work with. And South Africa is predominantly a white wine drinking nation."

He is reducing the area of his vineyards by removing the vines planted on the poorly-drained section of the vlei at the bottom of the farm. He will plant pasture for sheep grazing and will use a by-product as organic fertiliser to improve the quality of the vineyards on the slope.

Bonfoi wines are bottled and marketed by the Bergkelder in Stellenbosch.

Wine is not for sale on the Estate.

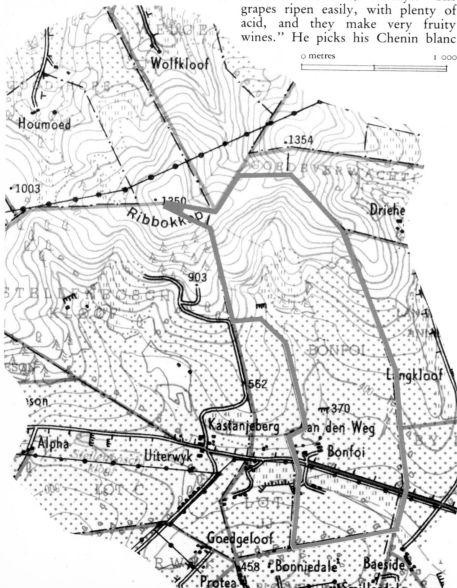

Delheim

One day Spatz Sperling climbed a ladder to the top of a fermentation tank. Within a few minutes he was so overcome by fumes that he fell back down the ladder to the floor. He hit his head on the way down and was out cold for several minutes. When he came to, he sat up and asked one of his employees, "What am I doing here?" The cellar assistant looked at Spatz for a moment and said, "Baas, we're making wine."

Michael 'Spatz' Sperling has been the driving force behind the Delheim Winery, situated on the Driesprong Estate, since 1951. The Estate was bought by Hans and Del Hoheisen in 1939. Hans Hoheisen was born in Worcester to parents who had immigrated to South Africa at the turn of the century. He became one of South Africa's pioneer Estate wine farmers, responsible for the development of Driesprong.

Michael Sperling arrived at Driesprong in 1951, having lost his family estate in Germany (now in Poland). As there was little future for a landless farmer in West Germany, he accepted the Hoheisens' invitation to join them in South Africa.

Delheim's cellars are situated on the Driesprong Estate, but they also handle grapes from the Vera Cruz farm. Delheim (i.e. Spatz and Hans) is the registered owner of both farms. Driesprong is 201 hectares in extent, of which 52 hectares are under vines. Vera Cruz is 135 hectares, with 76 hectares under vines.

When Hans Hoheisen bought Driesprong, the farm was covered with bush, and though it had no great history of successful wine growing, it was ideally situated to produce noble

The boundless energy of Michael 'Spatz' Sperling (left) and the dedication of Kevin Arnold have made Delheim a leading producer of semi-sweet and dry white wines.

The day the bird forgot how to fly, commemorated on a maturation cask.

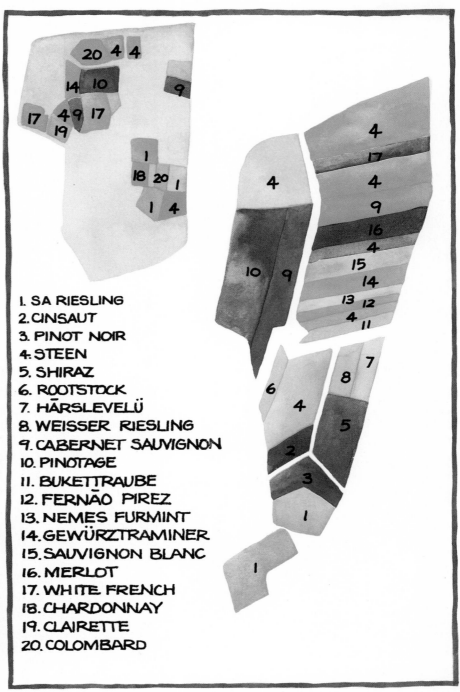

1. SA RIESLING
2. CINSAUT
3. PINOT NOIR
4. STEEN
5. SHIRAZ
6. ROOTSTOCK
7. HÁRSLEVELÜ
8. WEISSER RIESLING
9. CABERNET SAUVIGNON
10. PINOTAGE
11. BUKETTRAUBE
12. FERNÃO PIREZ
13. NEMES FURMINT
14. GEWÜRZTRAMINER
15. SAUVIGNON BLANC
16. MERLOT
17. WHITE FRENCH
18. CHARDONNAY
19. CLAIRETTE
20. COLOMBARD

Stellenbosch

wines. Driesprong is sited high on the south-east slopes of Simonsberg, immediately above Muratie, where Georg Paul Canitz was making great red wines, and between Uitkyk, home of the legendary Von Carlowitz wines and Schoongezicht, where noble Cabernet wines had been in production for nearly thirty years.

As Driesprong had never really been farmed until the advent of Hans, its early history is easily told. The first settler in the valley was Lorenz Campher, a German from Pomerania, who it is believed made his first home in the shadow of Simonsberg in the 1680s. In 1699 Willem Adriaan van der Stel granted him freehold title to 25 hectares of this land, which he called De Drie Sprong. Many years later this farm, together with the adjacent property Knorhoek, were added to the land holdings of the legendary Martin Melck, who already owned Uitkyk and Elsenburg, the latter now an agricultural college. The next owner was Jan David Beyers, Melck's son-in-law. He, and later his son, received extra grants of land extending the property further up the valley. Because Jan David Beyers rebuilt Campher's old home on De Drie Sprong from ruins, this property became known as Muratie, a corruption of *murasie*, meaning ruin. When the upper and lower farms were later separated, the higher farm retained the old name of De Drie Sprong, and until Hans's purchase little of importance happened to it.

During his first year on Driesprong, Hans planted Cape Riesling, Pinot noir, Cabernet Sauvignon and Gamay, with the intention of making wines of the quality standards set by his neighbours. In 1944 the vines were in full production and the cellar was built on the farm. This cellar, though extended and modernised, is still being used.

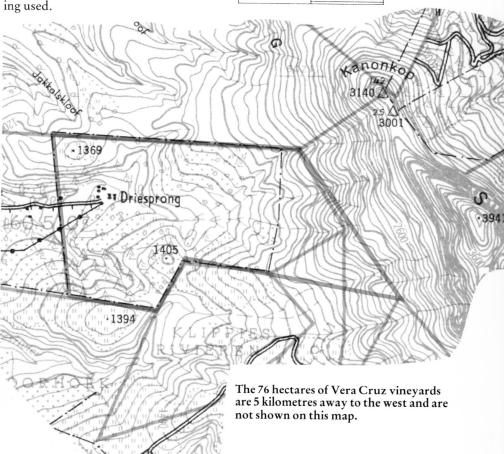

The 76 hectares of Vera Cruz vineyards are 5 kilometres away to the west and are not shown on this map.

Delheim's white grapes are being picked later, to make fuller-bodied white wines.

Delheim 135

The years from 1945 to 1965 were depression years in the South African wine industry. Brandy retained its position as the favourite drink of the nation and most of South Africa's wine growers were able to stay in production only by concentrating on supplying a large quantity of cheap inferior wine for distilling purposes. A few, and Hans Hoheisen was one, continued to produce high-quality wine, but received little reward for their efforts.

Spatz Sperling's early years on Driesprong were marked by a great deal of hard work and very little success in selling good-quality wine. These years were equally difficult for his neighbours. Schoongezicht was sold by the Nicholsons to the Barlows; Uitkyk also was sold by Mrs. Von Carlowitz to the Bouwers. In an effort to raise the revenue of Driesprong, Spatz cleared bush from steeply sloping land and planted large areas of pine forest. Even today, with Driesprong well established as a quality wine farm, it still contains the largest private pine forest in the Stellenbosch area.

During Spatz's first ten years on the farm, he was able to experiment and plan the areas of the farm best suited to each of the quality varieties already being used. He found that, as Driesprong is one of the highest farms in the Stellenbosch area, its soil and its micro-climate are very different from other Stellenbosch farms, even from those with boundaries adjacent to Driesprong.

Spatz has found that most red varieties, particularly the late ripening vines, have difficulty in fully ripening their grapes in the cool conditions provided by Driesprong. Hans Hoheisen's pioneer Cabernet Sauvignon, Pinot noir and Gamay vineyards and others planted later by Spatz struggled to reach ripeness and produced light wines without complexity or liveliness. But most white varieties seemed to appreciate the long ripening period, and some well-remembered Riesling wines were made in the early years from Driesprong's vineyards. This was a period when red vineyards brought fame and white vineyards brought income, and a wine cellar without a classic Cabernet in its range was believed to be insufficiently ambitious.

To provide extra potential, Spatz bought Vera Cruz, a farm approximately 200 metres lower in elevation, at the foot of Klapmutskop, and about five kilometres from the Delheim Winery. Vera Cruz receives twelve days more sunshine than Driesprong each year, and is well suited to producing full-bodied wines to stand alongside the delicate wines produced by Driesprong. Vera Cruz, about three kilometres further away from the mountain, receives about 30 per cent less rain than Driesprong.

At that time, Vera Cruz had mostly red vineyards. Most were Cinsaut and Pinotage, and a plan for redevelopment of the vineyards was drawn up. While the quality red and white varieties required for fuller wines were being planted on Vera Cruz, most of the vineyards on Driesprong were seen to be in need of renewal, and so a major programme of vineyard restocking has been under way for the past few years.

In 1980 Otto Hellmer, Delheim's cellar master for eight years, returned to his family's Estate in Germany's Rheinpfalz and the cellar responsibility was taken over by Otto's assistant, Kevin Arnold.

An important Delheim development has been the installation of small

Driesprong has Stellenbosch's largest private pine forest. The upper section was destroyed by fire in 1980.

Tucked away in the mountain, Driesprong's vineyards receive less sunlight than those planted in the valley.

Delheim's Driesprong vineyards are among the highest in the Stellenbosch area, and the cool micro-climate is best suited to early varieties.

These Riesling grapes are being crushed less than half an hour after being picked in the vineyard above the cellar.

oak casks for extensive maturation of red and white wines. Traditionally, most white wine has been made from early-picked grapes, with noticeably high acid content. These lively, fruity wines had been made for early consumption. But Kevin Arnold, benefiting from new vineyards of Weisser Riesling and Sauvignon blanc and an older vineyard of Gewürztraminer, has begun to make fuller wines. Today the vines are allowed to ripen the grapes further, and the chilled juice of the crushed grapes is allowed to remain in contact with the grape skins for longer periods to extract greater flavour. In some cases, the white musts are fermented in small oak casks to give the wine extra tannins for longer bottle life.

The semi-sweet wines that have drawn attention to Delheim in recent years have become a regular part of the programme. On Vera Cruz, *Botrytis cinerea* is found almost every year, and these noble rot grapes are used to make both Special Late Harvest and Noble Late Harvest wines.

The Delheim home and cellar complex is surrounded by vineyards on the lower section of the farm's Simonsberg slopes.

For the latter wine, the bunches are kept on the vines until the juice of their grapes averages about 40 ° Balling. When the bunches are brought into the cellar, they are carefully hand-selected for the most suitable material for both styles of wine. The grapes are crushed, the juice is chilled and the skins and juice are allowed to remain in contact for maximum flavour extraction. The must is fermented until the combination of sugar and alcohol is too much for the yeast to handle, and then it is cleaned and put into 225-litre oak casks for maturation before bottling.

Musts with lower sugar content, destined for the production of semi-sweet wines, have their fermentation arrested at the chosen degree of sweetness. The musts are passed through a centrifuge to remove the yeast and are then filtered sterile.

Grapes for the cellar's top red wines are allowed to develop a little extra sweetness before picking. The juice of the red varieties has traditionally been inoculated with dried yeast to start fermentation. Red musts containing the skins are fermented at temperatures of around 24 °C, as Delheim searches for richly coloured, lively red wines that are ready for consumption within a few years of the harvest. The skins are normally removed from the fermenting mash at about 12 ° Balling. The liquid must is then fermented out at cool temperatures in closed tanks. Once fermentation is complete and the wines have been cleaned, they are placed in small oak casks for a few months and then transferred to large wood for a final period of maturation before bottling.

In 1978 Otto Hellmer began to experiment with the production of port. Lacking specialist port varieties, he used Cinsaut, the only red variety he had available. He removed three quarters of the crop from the vines before the grapes were fully ripe and allowed the rest to reach a high level of sugar. The juice of these grapes was partly fermented before fortification with brandy and then matured in 300-litre oak casks. Port has been made in this style only when climatic conditions allowed it.

As an extension of the grape harvesting for Heerenwyn, Delheim's light summer wine, a quantity of low-alcohol Cape Riesling and Chenin Blanc wines were retained from the 1979 crop for the production of a tank-fermented sparkling wine. The base wines are stored in the tank for at least a year, then inoculated with yeast and fermented at a low temperature in a sealed vessel, which develops its own pressure. After fermentation, the wine is allowed to stay with the inactive yeast as long as possible. The wine is then filtered and bottled. The process takes more than two years from harvest to bottle.

Wine is for sale on the Estate.

Goede Hoop

The pebbly soils on the steeply sloping hillsides of the Goede Hoop Estate provide a harsh environment for the growth of vines. The vineyards are unirrigated and have to ripen their crop with the moisture they can extract from the clay base under the sandy loam topsoil. The vines are heavily pruned to restrict the size of the crop and allow the grapes to ripen fully, while retaining their natural acidity.

Compared with other Bottelary and Stellenbosch farms, Goede Hoop is relatively high and has vineyards on slopes mainly facing north and west in a rounded basin, sheltered from the summer south-east wind, where hot weather in the middle of summer can have a dramatic effect on sugar and acid content in the grapes.

Goede Hoop is the hilly section of what was once a much larger farm known as Welgevonden. The first known owners of Welgevonden were a family of Bosmans, who lived and farmed in the Bottelary district as far back as the seventeenth century.

Goede Hoop covers 122 hectares, of which approximately two-thirds are under vine. It is set far back from the Bottelary Road high on the side of the Bottelary range of hills, with every square metre of the farm on sloping hilly ground; the only parts of the farm not bearing vines are simply too steep to cultivate. The farm was separated from Welgevonden when an elder Bosman sectioned it off and granted it to one of his sons in the traditional manner of farming families during the late eighteenth century. Approximately two generations later the Goede Hoop section was sold to a member of the Krige family, and was to stay in their ownership for over a hundred years. During this long passage of time little wine was made on Goede Hoop, as it was believed that good wine could be made only on lower slopes and in rich alluvial soils, a belief that was to persist until well after 1900.

In 1928 Petrus Johannes Bestbier decided to farm in the Stellenbosch area and bought Goede Hoop from the Krige family. Petrus had grown up in Hopefield, near Saldanha Bay, where his family had been wheat farmers for generations. He had decided that wheat farming was not his game and was interested in developing vineyards. Up to that time Goede Hoop had been farmed mainly as a grazing property.

Petrus Bestbier saw great potential in the farm, quite apart from grazing cattle and growing small mixed crops. Most of the vines on the farm at that time were of unknown history and type. There was then a demand for sweet fortified wines, and Petrus planted a number of the vines in fashion at that time, for example White French, Green Grape and Steen. He later planted a quantity of Cinsaut vines because they were outstandingly durable, reliable and productive.

It was with Petrus's son, Johan, that the development of Goede Hoop as a quality wine farm began. Johan took over the management of the farm in 1961 after working for several years for his father. Dr. Niehaus of the KWV encouraged Johan to grow red cultivars for the production of port wines.

Johan began to cull many of the white varieties from the slopes of the farm and replaced them with Cinsaut, Shiraz, Tinta Barocca and Pinotage. Shiraz and Pinotage were originally planted for port wine production. He soon noticed that the red varieties adapted to the steep, gravelly slopes of the farm with alacrity. However, there was little demand for high quality red table wines at that time, and any further development of reds that he undertook during the 1960s was limited to greater production of Cinsaut.

With the expanding of the market for quality red and white wines in the early 1970s, Johan began to replan his farm. Though his particular interest was in the production of red table wines, the overwhelming preference of South African wine consumers was for white wines. To accommodate both income and interest, Johan maintained the existing proportion of white and red cultivars in his vineyards at 60 to 40 and converted his cellar to the exclusive production of red table wines. After many years of white port production, the white grapes were now sent to a wine merchant's fermentation cellar.

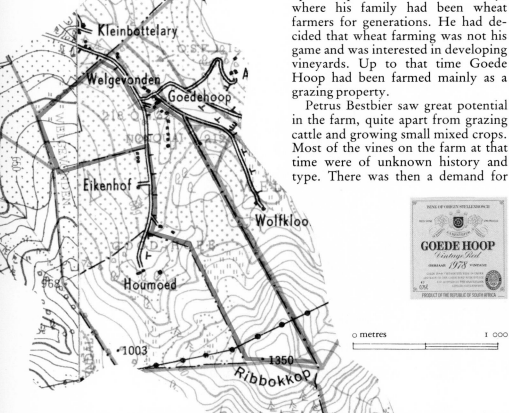

Johan Bestbier, producer of fine blended red wines, has been in charge of the vineyards and cellar since 1961.

The varieties originally planted for port production were the basis of the first Goede Hoop Vintage Reds, but in recent years Johan has established vineyards of Cabernet Sauvignon and Carignan and has extended the Shiraz vineyards. Shiraz is particularly well adapted to the Goede Hoop conditions, thriving in the broken, stony soil and the warm conditions and producing sweet, black grapes with sufficient acidity. These grapes are allowed to ripen fully and make wines with deep colour and a rich raisin-like flavour. As the vineyards of Goede Hoop are well suited to the production of full-bodied wines Johan allows all vineyards to ripen fully to make wines in this style. The juice of the red grapes is fermented together with the skins in open tanks at a moderate temperature, until most of the sugar has been converted. After the skins are removed, the wine is fermented dry in other tanks.

After fermentation, the wines are allowed to stand for a few months in the Estate cellar, before they are moved to a wine merchant's maturation cellar where they undergo a period of ageing in oak. Though the Vintage Red is a Shiraz-based blend, the proportions of the individual constituents change with each harvest and depend on the style and quality of the wines made from each cultivar.

Wine is not for sale on the Estate.

Goede Hoop, high in the Bottelary Hills, has low-fertility soils, and vines must grow without irrigation.

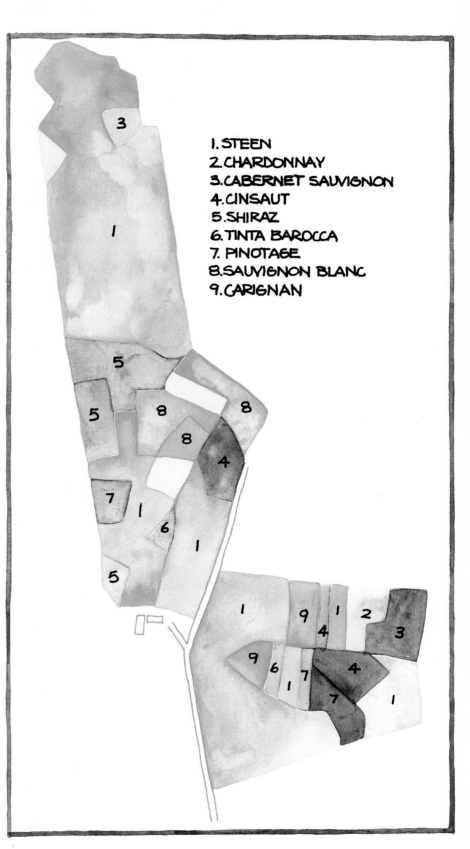

1. STEEN
2. CHARDONNAY
3. CABERNET SAUVIGNON
4. CINSAUT
5. SHIRAZ
6. TINTA BAROCCA
7. PINOTAGE
8. SAUVIGNON BLANC
9. CARIGNAN

Goede Hoop

Hazendal

Two of the major factors that have contributed to the success of the Hazendal wine Estate have been the degree of difficulty that the natural growing conditions have presented to the vines and the methods that Pieter Bosman has used to assist the vineyards to produce fine quality wines.

Located on the fringe of the viticultural area alongside the acidic, moisture-lacking soils of the Kuils River and Brackenfell areas, Hazendal has a multitude of different slopes and soil types that are unable to support large crops of grapes. Hazendal presents a collection of sometimes contradictory farming problems, which have been surmounted by care and logic and have repaid the patient approach by adding an indefinable extra element of quality to the flavour of the grapes.

The Estate is sited on the western end of the Bottelary Hills, and has slopes that face north, west and east. In addition there is low-lying level ground which in some cases is too sandy and deep-drained, and in other cases too wet and poorly-drained, to support vines. The soil types on most of the slopes are sandy and sandy loam, with patches of decomposed granitic and shale-based soils. In the vlei areas most soils are more sandy, and sometimes more fertile, than the farm's sloping ground.

During the eighteenth and nineteenth centuries, Bosman was the most common surname in the area generally known as Bottelary, and several of today's important wine Estates in this area were owned by members of that family. But only Hazendal remains a Bosman Estate, with the descendants of the vine-growing pioneers tending their hardy vines with matching determination.

Chenin blanc grapes develop rich flavour when grown on untrellised vines on Hazendal's sandy soils.

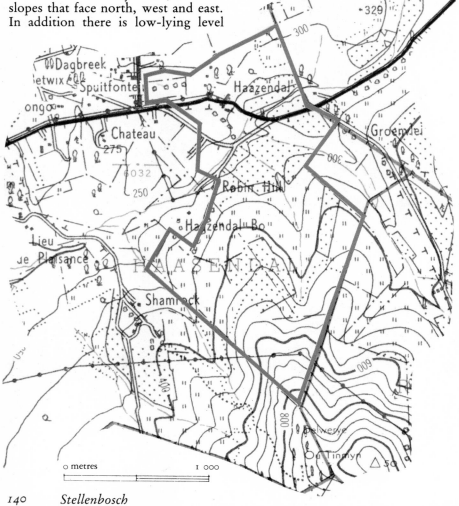

Hazendal's first owner was Christoffel Hazenwinkel, who came to the Cape in 1686 from Brunswick in Germany. In 1695 he became a Messenger of the Court of Justice and probably applied to be given burgher status and be allowed to take up property soon afterward. In the same year he married Margharita Michiels, daughter of the owner of the farm Joostenberg. It is not known when they were promised Hazendal, or when they moved to the property, but Hazenwinkel was officially granted the land in 1704. By 1712 he also owned the nearby Bottelary and Rosendal properties, but in 1718 he sold all his possessions in the Cape and returned to Europe. Hazenwinkel and successors used these farms as grazing land and only grew enough vines for domestic supply.

The Hazendal homestead, a superb example of the traditional style of Cape Dutch farmhouse, was built by the Van As family between 1770 and 1790, when Cape farmers encountered their first period of relative prosperity. Over the next two centuries, the condition of the structure deteriorated, but the building was restored by Pieter Bosman during the 1960s.

With poor soils, Hazendal did not prosper during the boom periods of the Cape wine industry, when emphasis was firmly placed on greater quantity of production. Pieter Bosman's father, Isaac, had to restrict the

The late Pieter Bosman restored the Hazendal home during the early 1960s.

Michael Bosman is the latest in a long line of Bosmans to grow vineyards along the Bottelary.

Hazendal is a fringe farm. The sandy soil in the foreground constitutes part of the western edge of the Stellenbosch winelands.

growth of his Cinsaut, Palomino, Sémillon and Chenin blanc vines to ensure a reliable crop and made broad, full-bodied wines that attracted the attention of the KWV wine experts when the market for sherry and port developed during the first half of the present century. Isaac made these wines in Hazendal's simple farm cellar, with the port wine partly fermented in open concrete tanks before fortification with brandy, and the sherries fermented dry, before wine spirit and flor yeast were added to provide the required character.

Pieter Bosman joined the farm in 1941 and continued along the course developed by his father, with small yields of sweet grapes grown on bush vines, exposed against the hillsides to the morning or afternoon sun. Under these conditions, lesser quality varieties were able to concentrate their sweetness and flavour to make notable wines.

As the port and sherry market declined during the 1950s, the South African public began to take an interest in table wines for the first time, and the more adventurous producers installed cold fermentation equipment to be able to bring the conditions under which the wines were made closer to those existing in European cellars. Faced with the insecurity of this market, but aware of the necessity to prepare for change, Pieter Bosman rebuilt the Hazendal cellar, beginning with the installation of crushing, chilling and storage facilities, and temporarily halted the fermentation of wines, concentrating on making top quality grape juice for sale to wine merchants for fermentation in their own cellars. Pieter's refrigerated grape juice cellar was first used to process the 1964 crop and for another nine seasons no wine was made on the Hazendal Estate. During this period Pieter replaced most of the Sémillon and Palomino vines with more Chenin blanc and added Pinotage and Tinta Barocca to the red vineyards.

By 1973 Pieter had installed enough stainless steel tanks in his cellar to be able to ferment his first dry table wines, which were then sold to a merchant. The quality of the Chenin Blanc wine from Hazendal made an impression, and the 'guava' flavour that is obtained from fully ripe grapes on certain farms was seen to occur each year in the tanks of wine made from this cultivar in the Estate cellar. In 1974 a quantity of Chenin Blanc was bottled (using the South African name Steen).

Pieter planted Chenin blanc vines on each of the slopes of the farm as well as in the deep sandy soils, both dry and wet. He came to the conclusion that Chenin blanc produces its best wines when the vines are grown as bushes on the slopes facing the morning sun. The vineyards in the fertile, flat ground had to be trellised to support the more vigorous growth and were found to produce wine of lesser quality. Each year the Chenin blanc grapes were picked at mid-ripe stage (21-22 ° Balling) and fermented dry. A small quantity of Chenin Blanc grape juice was retained to add

Hazendal

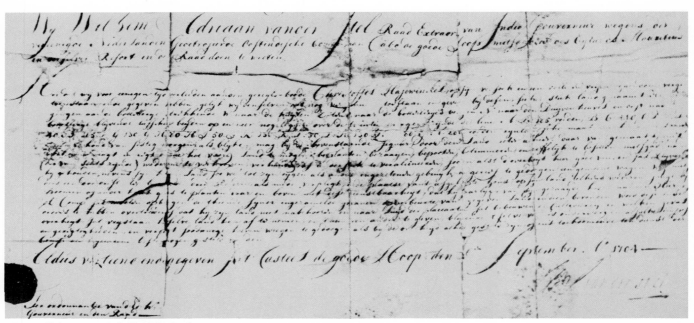

Christoffel Hazenwinkel's title deed now belongs to the Bosman family. The Bosman crest has had pride of place in the Bosman home since the early 1800s.

to the dry wines, to produce a semi-sweet product.

At the end of the '70s Pieter was joined by his sons Michael and Idee, who were able to assist with the development of new vineyards and cellar expansion. Vineyards of Shiraz and Cabernet Sauvignon planted by Pieter had come into production, new vineyards of Rhine Riesling and Sauvignon blanc were planted, and vineyards of Pinotage were removed.

By 1982, when Pieter died, Hazendal's 147 hectares contained 80 hectares of vineyards in roughly equal quantities of red and white varieties. Beyond assistance provided to young vines, none of the vineyards is irrigated after maturity. During a particularly warm or dry summer Idee Bosman's team takes a water cart through each of the young vineyards, and after having made a depression around each vine, adds water to help the plant survive until the winter rains.

Having benefited from several seasons with his father in the Estate cellar, Michael took over the production of wine on Pieter's death. In 1982 the first late harvest wines on the Estate were made from very ripe Chenin blanc grapes. These grapes were allowed to reach high sugar levels, and the acidity of the grape juice was adjusted to improve the balance of the wines before fermentation. The sugar content of the musts was reduced, in the normal manner, at cool temperatures until the liquid has reached the desired late harvest degree of sweetness, when the fermentation was arrested by lowering the temperature and raising the SO_2 content. These full-bodied, semi-sweet wines were stored for a short period in stainless steel on the Estate before transportation to a Stellenbosch cellar.

Red wines from Cabernet Sauvignon, Shiraz and Cinsaut are made in the Estate cellar. In most years, Pieter made lighter, medium-bodied reds and Michael is following his father's lead. The juice is fermented with the crushed grape skins until two-thirds of the sugar has been converted, and from this point the juice is fermented dry on its own. These wines are produced for blending by a wine merchant. There is no immediate prospect of a Hazendal Estate red wine, though Michael, like his father, believes that the farm is best suited to the production of red wines.

Full-bodied, semi-sweet Chenin Blanc wines with a rewarding capacity to improve with a few years' maturation will continue to represent the Hazendal range.

Wine is not for sale on the Estate.

Jacobsdal

Jacobsdal lies on the southern edge of South Africa's wine-producing country, facing False Bay. The farm has no water other than the 600 mm that is the average winter rainfall in the area, and has such sandy soil that vines, with their deep-penetrating root structure, are probably the only crop the farm could support.

The grape varieties have to be chosen for their strength and durability, as weak-growing and delicate vines would be unable to ripen a healthy crop of grapes under Jacobsdal's difficult conditions. The hardy grape vines have to be carefully tended to ensure the growing of new wood and the production of grapes each year. On the soils of Jacobsdal, a vine, if allowed to bear unchecked, is capable of wearing itself out within a few years. Consequently, the growth of all vines is cropped back to provide only the minimum necessary for survival through the winter, and to be able to bear a crop in the spring and summer. Many of the developing bunches are pruned off to allow the remaining ones to utilise the plant food and moisture provided by the extensive, deep-reaching root system. As a result, every vine on Jacobsdal concentrates its efforts into producing a small number of grapes. This has been standard procedure on the Estate since Cornelis Dumas bought Jacobsdal in the early 1920s.

Many of the Jacobsdal vineyards have rows 1,3 metres wide and have to be cultivated by horse and hand.

Jacobsdal Estate covers 500 hectares of land, but most of this land is low-lying, flat ground, almost pure sand, which is unsuitable for vineyards. The farm has about 120 hectares of sloping land from 50 metres to 150 metres above sea level, best suited to drought-resistant, strong-growing cultivars. Cornelis Dumas the elder — another Cornelis Dumas is the present owner of Jacobsdal — recognised this potential and, believing in the affinity of red cultivars and sea breezes, chose to plant Cinsaut as well as Steen. Some of the vines on Jacobsdal today are over sixty years old and were on the farm when Cornelis bought it. Because the vines he planted did not have to work hard to produce to capacity, many of them are producing as well today as in their prime, which is often considered to be within the first twelve years. From these Cinsaut and Steen vines Cornelis made half-sweet vines, by fermenting the wine down to about 10° Balling and then fortifying it with wine spirit. These wines were very popular in his day, for all cultivars grown on Jacobsdal are able to develop a high sugar content before harvesting, important for this style of product.

Cornelis Dumas, specialist in full-bodied Pinotage wines.

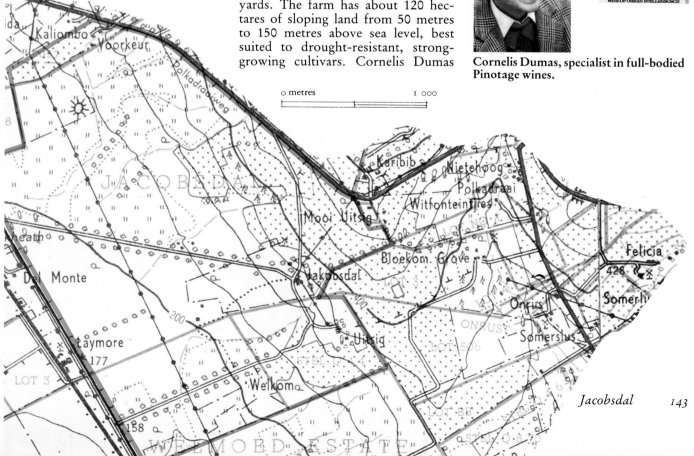

Cornelis's son Petrus Johannes (Hansie, as he was generally known) took over the farm after his father's death in 1951. At this time, the proportion of vines in the 85 hectares of vineyards was about 85 per cent Cinsaut to 15 per cent Steen. Hansie Dumas extended the Cinsaut vineyards further along the slopes and began to plant Pinotage, believing that what had proved successful with Cinsaut would work equally well with its descendant. He began to make dry red wines from Cinsaut and Pinotage, and dry white wine from Steen.

In 1966 Hansie Dumas died, and his eldest son Cornelis returned from his studies at Stellenbosch University to take over the farm. Cornelis has followed his father's practice and has replanted some of the older Cinsaut and developed new Pinotage vineyards to the stage where the farm now has about 100 hectares under vines.

At present, Cornelis has approximately equal areas of Pinotage, Cinsaut and Steen, with smaller blocks of Sauvignon blanc and Cabernet Sauvignon. All the vines are grown as bushes, though the Estate will slightly modify its policy to provide a low trellising system to tame the wild growth of the small Cabernet vineyard. All vineyards are grown without irrigation on sloping ground that faces south-west toward False Bay and receives afternoon sea breezes throughout the summer ripening season. In addition, the humidity brought in from the coast occasionally forms evening mists and provides regular deposits of dew during the warmest part of the year. The growing conditions in the elevated vineyards contribute to the individuality of the Jacobsdal wines.

Jacobsdal's white varieties are grown only to keep the harvesting team busy while waiting for the Cinsaut and Cabernet Sauvignon grapes to ripen.

The Estate is best known for the production of an easy-drinking Pinotage wine with a most individual style. The vines are planted about one metre apart and grown without additional water in the traditional low bush shape, without support.

These vines are capable of reaching very high sugar levels, but are normally harvested full-ripe (22-23 ° Balling) to make full-bodied wines. After crushing, the grape skins remain undisturbed on top of the grape juice in the open concrete fermentation tanks, until the yeast in the vineyards and cellar begins to ferment the sugar in the juice. Once the first few tanks of Pinotage start fermenting in this manner, must with active yeast is added to each tank to hasten the process. After the yeast has begun to reduce the sugar, CO_2 is produced, protecting the must from oxidation. The cap of compacted skins is broken up and mixed with the must to obtain the desired colour.

Cornelis Dumas is dedicated to the job of making soft but full Pinotage wines. He examines the colour and evaluates the flavour of the must as it ferments to estimate the ideal separation point, and removes the skins from the mash, which has been cooled and maintained at a reasonably constant 24 °C, before tannic flavours are imparted. The must is fermented in open tanks until most of the sugar has been removed, and is then transferred to closed tanks to complete the process. The dry wines are racked to remove solid matter and remain in concrete tanks in the Estate cellar for several months to allow the desired malolactic fermentation to take place. It is then transported to a Stellenbosch cellar for maturation in oak.

Although the Estate's Cinsaut and Cabernet Sauvignon grapes are crushed and fermented in the Jacobsdal cellar, there is little likelihood of these cultivars being marketed under the Jacobsdal Estate label in the immediate future.

Wine is not for sale on the Estate.

On Jacobsdal's sandy soil, Pinotage vines are pruned back to reduce the crop of this fertile variety.

Because of the poor soils, production is controlled, and vines are grown as bushes without trellising, on Jacobsdal.

Fewer than 10 kilometres from the sea, on higher sloping ground, Jacobsdal receives the cooling breezes of summer.

Kanonkop

Looking over the first shoots of spring towards the Estate's new white variety vineyards, high on the Kanonkop foothills.

The Estate's mixed red vineyard is in the foreground and Pinotage and Cabernet Sauvignon are on the slope.

Stellenbosch's early-morning mists reduce the hours of direct sun and increase the humidity.

When the Sauer family sold the manorial Uitkyk Estate to George von Carlowitz in 1930, a 90 hectare portion on the lower slopes of the farm was retained to provide Paul Sauer with a farming interest. Most of the vineyards on the Uitkyk Estate had been on the land nearer the valley floor, and so this lower farm, which they named Kanonkop after the signal point on the peak above Uitkyk, had established vineyards but no cellar or homestead.

The soundness of Paul Sauer's choice of the warmer farm has only been appreciated during the last few years. The belief that cooler slopes produce the best South African wines was firmly held during the '60s and '70s, but in most instances, where clear comparisons are available, factors other than temperature alone seem to have greater importance. At Uitkyk, Cabernet Sauvignon vineyards planted on the elevated slopes have proved to be less than ideally placed and have been removed. On Kanonkop, where the difference between the highest and lowest vineyards is 130 metres, Cabernet Sauvignon is grown on cooler ground on the middle slopes as well as on low level warm soils — and the finer wines have been made from the produce of the lower vineyards.

Paul Sauer planted Chenin blanc, Sémillon, Clairette blanche and Cinsaut during the early years. When the production of table wines began to cause interest in the market-place, Paul was one of the first farmers to plant Pinotage, and this variety was to have a great deal of influence on the course of events in the South African wine industry as well as on Kanonkop. The vines were released for commercial planting in the mid-1950s and Paul was one of the first two producers to plant a vineyard, placing the vines in warm soil on the crest of a low ridge. One of the first crops of grapes produced wine that won the red varietal wine championship at the Cape Wine Show in 1961. To date, Kanonkop is the only wine cellar to have defeated the Cabernets with a Pinotage. The performance was repeated at Stellenbosch in 1982, with wine made from an adjoining vineyard, grown under similar conditions. Following Kanonkop's initial success, other producers were quick to plant this new variety that could produce a large crop of grapes and make champion wine. So while his colleagues planted Pinotage, Paul planted Cabernet Sauvignon, also on the lower ground.

Paul Sauer, the son of Senator J. H. Sauer, followed his father into politics and entered Parliament in 1929. Father and son spent a total of 82 years in the House of Assembly and the Senate and each spent time in the Cabinet. At various times, Paul Sauer held the portfolios of Transport,

Lands, Water Affairs, Forestry and Public Affairs. He was the only member of the wine industry in Parliament and publicly encouraged the drinking of 'common or garden, light, wholesome, honest dry wines for everyday use'.

Paul Sauer had two daughters and was without an obvious heir to continue the tradition on Kanonkop. In 1968 he asked his daughter Mary and her husband Jannie Krige to take over the management of the property. Jannie, who was the Sports Administrator at Stellenbosch University at the time, decided to appoint a final-year viticultural student, Jan Coetzee, as farm manager. It was only in 1978 that Jannie finally joined the manager Jan, known generally as Jan Boland, on the farm. By this time Kanonkop Cabernet Sauvignon and Pinotage wines had been on the market under the Estate label for several years. The Cabernet and Pinotage vineyards had been extended to cover most of the sloping ground on the lower section of the farm, and further vineyards of these varieties, together with Pinot noir, had been planted on the high slopes. Chenin blanc, Colombard and Cape Riesling had been planted on the level ground between the cellar and the main road, except for a wet section that is used for pastures.

A second farm, Kriekbult, a few kilometres away from Kanonkop, had been bought by Paul Sauer in 1940, and was incorporated into the Estate. The sandy loam and shale-based soils on Kriekbult's 60 hectares were well suited to vineyards, but the Estate management has always believed that better wines are produced from the crops of the home farm vineyards.

Jan Coetzee established a viticultural style on Kanonkop that allows each cultivar to produce to its greatest potential. He was one of the pioneers of the thorough preparation of soils to allow optimum root development for vines. Large quantities of lime were worked into Kanonkop's calcium-deficient soils and the chemical analysis of the soil was adjusted to ensure vigorous growth. In the late '70s Jan started to plant Weisser Riesling and Sauvignon blanc to improve the quality of the white wines being made on the farm, and aimed to eventually replace most of the standard white varieties with more noble cultivars. At this point Kanonkop had not produced any outstanding white wines, although 45 per cent of the vineyards were white. Jan also planted a vineyard of Cabernet franc, Merlot, Malbec, Souzão and Cabernet Sauvignon, the intention being that these grapes would be picked and crushed together, to make Kanonkop's first high-quality blended red wine.

In 1980 Jan Coetzee left the team on Kanonkop to try to realise his ambition of making great wines on a small scale on his own property. Jannie chose Beyers Truter, another Stellenbosch graduate, to take his place. Beyers, with a limited amount of experience in a co-operative cellar and in the production of table grapes, brought a methodical approach to the Kanonkop vineyards and cellar. He found the Estate's red-blend vineyard and Weisser Riesling ready to produce their first sizeable crops and learnt that the vineyards also produce fine-quality grapes in cool seasons with difficult ripening conditions. With rain during the 1981 harvest and generally cooler weather, the grapes from the Cabernet Sauvignon vineyards were picked mid-ripe (21-22 ° Balling) and made a wine that won the Stellenbosch red wine Championship at the annual young-wine show.

There are three main soil types on Kanonkop, with most of the vineyards planted in red Hutton soils. On the lower slopes this soil is loamy, and in the higher vineyards the texture is pebbly. Both these soils contain clay and retain high proportions of moisture throughout summer. The red soils and the sandy loam soils are based on a layer of clay that forms a water table.

An additional 46 hectares of land have been added to the higher portion of the farm, increasing the area of vineyards growing under cool conditions. Weisser Riesling, Sauvignon blanc and Cape Riesling have been planted on a wind-swept crest that offers north-, east- and south-facing slopes. An area on the very top of the ridge has been reserved for Chardonnay. The warm growing conditions in the lower vineyards and the availability of natural water to the vine roots right through the ripening season, combined with the Estate policy of small crop expectations from deep-rooted vines, have ensured that Kanonkop has made mostly full-bodied wines of considerable distinction. The Pinot noir vineyard, mid-way up the slope, and the more recent white vineyards in their high positions will pro-

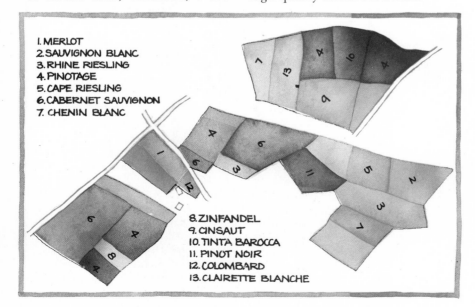

Beyers Truter, who landed in Kanonkop's hot seat in 1981 and made championship-winning wines.

The first signs of new growth can be seen on Kanonkop's early-ripening varieties in mid-September.

vide the farm with a further perspective in wine styles. The Pinot noir has already shown promise and has made fruity, medium-bodied wines, light in colour but complex in flavour.

The red wines are fermented in open concrete tanks, using dry yeast to start activity in the mash of juice and skins. Beyers uses the degree of sweetness in the grape juice produced by the warmth of the ripening season to decide the style that the fermentation will follow. He maintains the fermenting red must at relatively cool temperatures (22-23 °C) until he judges that the liquid has gained the tannic extracts and colour he requires (more in a cool year, less in a warm year), and then he presses the remaining liquid from the skins. Some of this pressed juice is added to the must that has now been transferred to closed stainless steel tanks. The wine ferments dry, is encouraged to undergo a malolactic fermentation, and when six to eight months old is put in oak for at least one year to gain complexity. Pinot Noir, unlike the other reds, is matured in small oak casks for a shorter period.

The juice of Kanonkop's white grapes is fermented at cold temperatures (around 12 °C) until dry. These wines are used to produce a blended Bouwland Wit (Chenin Blanc-Weisser Riesling) and pure cultivar wines.

At present about 20 per cent of Kanonkop's production is Estate-bottled and the balance is sold to a wine merchant in Stellenbosch.

Wine is for sale on the Estate.

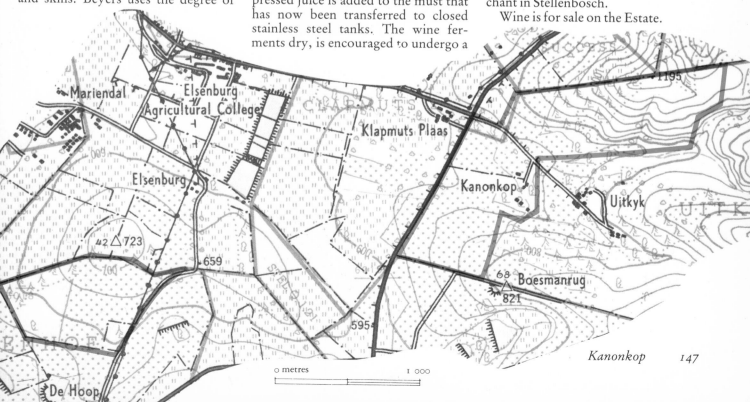

Koopmanskloof

The Koopmanskloof cellar and home are among the vineyards on a north-facing slope in the Bottelary Hills.

(below) Chenin blanc vines are grown as bushes, without irrigation, for low yields on the Estate's sandy soils.

Stewie Smit, producer of quality grapes on low-fertility soils.

Blanc de Marbonne is a light, dry and refreshing wine with a flowery style, blended from the produce of Chenin blanc, Colombard and Clairette blanche grapes grown on South Africa's largest wine Estate, Koopmanskloof. The story of the operation of this vast Estate is pertinent to the background of Blanc de Marbonne, as the Chenin blanc component that dominates the blend may have been produced in any of many different sections of the Estate.

Koopmanskloof is first mentioned in historical documents as an 18 hectare portion of the farm Bottelary. It was sold in 1787 to Petrus Bosman, described as a tilemaker. Many Stellenbosch soils, including those of Bottelary, have a high proportion of clay and include pure clay deposits. Bosman used the clay to produce baked stone tiles for floors and ceilings. The floor of the Strooidak Kerk in Paarl has tiles made by the first owner of Koopmanskloof. (Today these tiles are covered by carpet.) In 1801 Petrus built a gabled house on land that adjoined his farm, land that he was undoubtedly farming, even though he did not own it until 1818. Wynand Smit, who had previously farmed on the Cape flats, bought Koopmanskloof from Petrus Bosman Jnr. in 1896, and the farm has remained in the Smit family since then.

Koopmanskloof is mostly north-west-facing and the vineyards are grown on sandy, loamy and stony soils between Koelenhof and Kuils River. The vineyards climb the slopes of the eroded Bottelary Mountain, providing a certain amount of variation in elevation for the selection of cultivar and vineyard.

Stewie Smit, the present owner of the Koopmanskloof Estate and grandson of Wynand Smit, inherited the bush vine vineyards of the home farm and set out to get the most from the restricted elements with which he had to work. He found that Chenin blanc grew reasonably well and made most acceptable wines under the harsh conditions the farm presented. His mastery of the difficulties of Bottelary farming and his success with Chenin blanc vines led to expansion and the Estate now contains five adjoining farms, with a separate farm, Watergang, on the north-west boundary of Stellenbosch. With the exception of Watergang, the farms that make up Koopmanskloof share similarly infertile soil and a lack of summer water. Chenin blanc is grown on each farm and in almost all cases the vineyards contain bush vines.

The grapes produced by the vineyards grown on the sandiest soils have

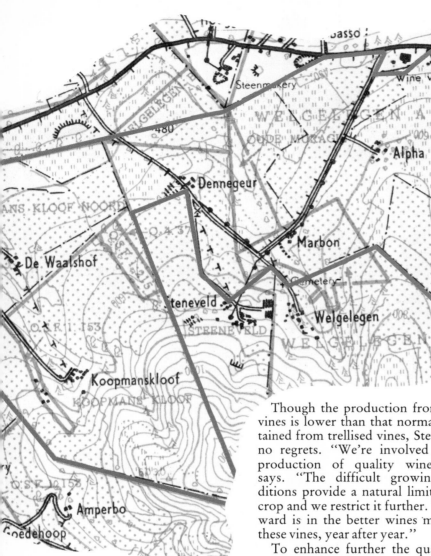

The sandy texture of the Koopmanskloof soils may be seen as these Chenin blanc cuttings are planted.

trouble with lack of moisture as the summer wears on, and as they reach mid-ripe stage with plenty of acidity early in the season, Stewie picks these vineyards first. It is from these fruity, early-season musts that the wines used for Blanc de Marbonne are generally chosen. To extend the season and allow the cellar to ferment the large volume of Chenin blanc juice, those blocks with sandy loam, shale-based and clay-mixed soils are left to the middle and even the end of the season to harvest. Though we are unlikely to taste these fuller wines under the Koopmanskloof label, the 600 hectares of Estate vineyards produce a very large volume of Late Harvest and Special Late Harvest wines.

Though the production from bush vines is lower than that normally obtained from trellised vines, Stewie has no regrets. "We're involved in the production of quality wines," he says. "The difficult growing conditions provide a natural limit to the crop and we restrict it further. The reward is in the better wines made by these vines, year after year."

To enhance further the quality of Koopmanskloof's production, the vineyards planted since 1979 have been provided with pairs of narrow rows separated by a wider service row, increasing the vine population in each block. The block of vines can produce more than a similar block with widely separated rows, but each vine will be expected to produce a smaller quantity and, in a healthy vineyard, this will automatically assist in the improvement of quality.

In contrast to the main section of the Estate and its predominantly Chenin blanc vineyards, Watergang has Hutton-type red soils with clay content. Here Stewie has planted varieties that require trellising for optimum performance and supplies them with irrigation from a farm dam. He has planted Rhine Riesling and Sauvignon blanc, which may be used as bases for future Koopmanskloof wines, and Cabernet Sauvignon and Shiraz. Pinotage, Tinta Barocca and Carignan are planted in sandy soils and are grown as bushes. "There's a place for Carignan in poor, sandy soils," says Stewie. "It's such a healthy grower and vigorous bearer, I do believe that by pruning it back to obtain a moderate crop and letting it struggle under our conditions we'll get very good quality light red wines, year after year."

The maintenance of such a large vine population, the collection of a major grape harvest in good condition over two months and the conversion of the juice of these grapes into quality table wines, requires a well-motivated and organised team. Stewie has a motivational programme that involves his farm managers, and through them the teams in the vineyards, where they evaluate the techniques and progress in the Estate programme. The cellar work and wine making also involve teamwork as Koopmanskloof is now a member of the Bergkelder's programme that provides cellar assistance and wine marketing for Estates.

During the early '70s Koopmanskloof bottled a small proportion of the cellar production for sale, but now the maturation and bottling tasks are the responsibility of the staff of the Bergkelder.

Wine is not for sale on the Estate.

Le Bonheur

"I think soil is probably the most important factor in the making of quality wine," says Michael Woodhead, vintner on the Le Bonheur Estate. "If you plant the most wonderful variety in a soil that is inadequate, you're going to make inferior wine. We're blessed with some very poor soils in this area, particularly low in mineral content, and before we can possibly think of making quality wine, we have to bring the soils up to par."

Michael Woodhead is the son of a South African champion motorcycle rider who died in a track accident when the boy was five years old. After spending his childhood in Pretoria and Natal, Michael went to Holland to study soil science and tropical agriculture. He has worked as a land planning officer in Tanzania, as a forestry officer in Canada, and has owned small stock farms in Swaziland and Knysna, where a severely broken leg caused him to look for a farm with less emphasis on physical activity. He chose to work with vines and bought a derelict property with vineyards and cellar on the border between Stellenbosch and Paarl.

When Michael came to Le Bonheur in 1971 he had had no experience of viticulture or wine production, and during the initial development of the farm acted on the advice of colleagues and professional advisers. Today he has the advantage of over a decade of experience to illuminate the course of the Estate he owns in partnership with the Bergkelder.

Le Bonheur has 60 hectares of vineyards, and a further ten are being prepared for planting. It takes Michael about three years to prepare the soil to a condition ready for the healthy establishment of a vineyard. "Our soils are deficient in calcium, magnesium and phosphate," Michael says. "We have to add up to 20 tonnes of lime to get the calcium to an acceptable level, and if you add all of that at once, you'll end up with very unbalanced soil. I also have another theory. A vine is a forest plant. The American ancestors of our rootstock varieties grow naturally in humus-rich soils in forests and I try to get as much humus as possible into the soil while we're raising the mineral content."

Each year Michael grows a winter cover crop in the soil that he prepares for vines, and, if the land can be irrigated, also plants a cover crop in summer. This lush vegetation is ploughed into the soil and thoroughly mixed. He adds over 100 tonnes of organic material, most of it municipal compost, to each hectare before he tills the soil for planting.

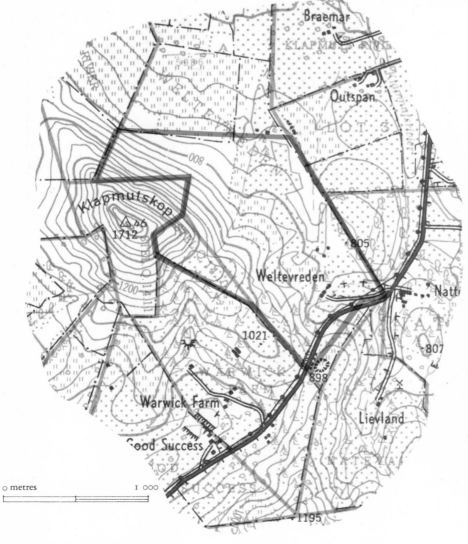

With 200-metres variation between highest and lowest vineyards, grapes ripen from February through to April.

Stellenbosch

Le Bonheur lies at the western end of the Simonsberg chain of Estates, sharing soil types and summer breezes.

Running from the base to the peak of Klapmutskop, Le Bonheur's slopes face north, south, east and west.

The Le Bonheur Estate was originally part of the farm Natte Vallei, granted in 1715 to the ex-soldier Jurgen Hanekom, who arrived at the Cape in 1708 and was indentured to the Company as a wood-cutter. He was promised land in 'De Witte Vallei' the following year and was given permission to graze cattle there in 1713. Soon after he was promoted to the rank of corporal and granted freehold ownership of the 34 hectare farm. Natte Vallei later became the property of the De Villiers family, and when the British administration instituted the permanent quitrent system of land ownership, Abraham de Villiers obtained 650 hectares of adjoining land in 1819. The south-western section of this land, from the lower slopes to the peak of Klapmuts Kop, the last link in the chain of the Simonsberg Hills, was separated from Natte Vallei in 1820 and given to Jacob de Villiers, Abraham's brother. This farm was known for 160 years firstly as Weltevreden and later as Oude Weltevreden, before the possibility of a confusion with the Bonnievale wine Estate, Weltevrede, caused the change to Le Bonheur.

There is a 200-metre difference in altitude between the most elevated vineyards on Le Bonheur and those in the base of the valley, and though most of the mountainous farm has west- and north-facing slopes, there are also east- and south-facing vineyards over the crest of the ridge. The Estate has five basic soil types, beginning on the higher section of the farm with light red granitic soil and a heavy red clay soil, changing to yellow sandy clay coming down the slopes and varying from sandstone derived soils to pot clay, most often with a shallow sandy loam overlayer. All these soils share the original mineral deficiencies that Michael is trying to correct, but each of them presents unique problems in choice of grape cultivar and rootstock partner, to achieve the desired result of a vine with vigorous growth and healthy, well-balanced grapes.

Because a substantial area of the farm has naturally acid, pot-clay soil, either on the surface or below a layer of sand, Michael began to develop this land for the planting of noble varieties. He started by ripping the clay and sand to a depth of over a metre during early spring when the clay was soft and then adding the minerals and humus required. By 1978 he had an area of pot clay, normally left for pastures, suitable for vineyard establishment, and planted the Estate's first Sauvignon blanc vines. A noble variety, Sauvignon blanc was chosen for its ability to grow vigorously under difficult conditions.

"It grew well, almost embarrassingly well under those conditions," says Michael. "And so we chose to plant Sauvignon blanc on most of our cooler clay soils. We plant cover crops between the rows in summer and plough them in, but the vineyards seem to have been so well prepared that we don't need to add fertiliser at all now, and we're getting respectable crops of up to twelve tonnes per hectare, with grapes that have plenty of sugar and acidity. What's more, the acidity remains in the wine. On low pH soils, with low calcium content, grape musts can lose as much as half their acidity during fermentation. On our adjusted soils, Sauvignon blanc musts lose only 1½ gℓ during fermentation, and the wines are naturally balanced."

Michael has chosen to grow only Chardonnay and Sauvignon blanc vines for the production of white wine. The Estate has Chenin blanc vineyards that were planted by the previous owner, and these will be replaced by future plantings of the two

Le Bonheur

classic whites. The Chardonnay vineyards have been planted in areas with well-drained soils and maximum exposure to the sun. Michael's plan is to have the Chardonnay grapes fully ripe and picked before the maximum heat periods of February arrive. He has found that the warmth of January brings Chardonnay to full ripeness with plenty of acidity. The grapes are picked mid-ripe (21-22 ° Balling) and the stems of the bunches are removed as the grapes are crushed. The juice and skins remain together in the drainer as it is filled, a job that normally takes four to five hours, providing for a period of juice and skin contact. The freely available liquid is run off, then the remainder is pressed in the basket press to obtain every last drop of juice to add to the fermenting must. In common with many other wine producers, Michael has found (contrary to his previous experience with Chenin blanc) that the more noble varieties can be pressed heavily, and all freely available liquid can be used to make quality wine. The Sauvignon blanc juice is fermented at a moderate 16-17 °C, while the Chardonnay juice is fermented at a higher temperature (20 °C +) until all the sugar has been removed. The Chardonnay wine is cleaned and placed in oak for maturation, while the new Sauvignon Blanc wine is retained in closed stainless steel.

The oldest quality-variety vineyard on Le Bonheur has been the source of many problems, and yet has been responsible for the production of one the Estate's best-known wines. Cabernet Sauvignon vines were unfortunately grafted on to a 101-14 rootstock, best suited to an early-ripening scion variety, and planted on a cool, high, south-facing slope in 1972. Each year the grapes of the late-ripening Cabernet were approaching ripeness just after change of colour, when the rootstock began to shut down for the winter. The result was a small crop of barely ripe grapes from a variety that is believed to require full ripeness to obtain the desired depth of colour and flavour. In the process of pumping juice back into the crusher to enable the pump to move the mash, the juice obtains a rosé colour, and being relatively low in sugar content, and moderately high in acidity, is ideally composed to produce a fresh, light rosé, suited to early consumption.

So far, the red vineyards have not been as successful as the white. Michael Woodhead has found that Shiraz does not produce its best wines on the warm, dry slopes of the Estate, and Carignan vines have recently been planted in the deepest of the sandy soils to make best use of this variety's ability to thrive and produce well under adverse conditions.

The red, white and rosé wines of Le Bonheur are transferred to the Bergkelder's cellars for maturation and bottling.

Wine is not for sale on the Estate.

Though high vineyards cause problems for Cabernet Sauvignon ripening, cooler conditions benefit white varieties.

The lower Estate vineyards have clay and sandy loam soils which have been reconstituted.

The Le Bonheur house and cellar buildings date from the early nineteenth century.

White wines made from Sauvignon blanc and Chardonnay, and a Cabernet Sauvignon rosé, are made in this cellar.

Lievland

Hidden from view, even from the highway that passes the farm, Lievland is wholly contained within a horseshoe-shaped valley, carved out of the lowest ridge in the Simonsberg, on the shoulder of the Kanonkop. Lievland forms the second unit in a chain of important Estates that share boundaries along the southern slope of the Simonsberg, stretching from Le Bonheur to Schoongezicht. Each of these farms has higher rainfall (750 mm) than the Stellenbosch average, decomposed granite and clay-mixed Hutton soils, and shorter hours of sunlight than the farms in the valley. They also benefit from greater exposure to the cooling effects of summer breezes, though Lievland's sheltered position in a basin probably reduces the exposure to southern winds. As the shallow valley faces west, Lievland has vineyards on south-facing, west-facing and north-facing slopes that provide the Estate with a wide range of growing and ripening conditions and enable the cellar to pro-

The basin of the Lievland Estate has slopes that face south, west and north, providing varied growing conditions.

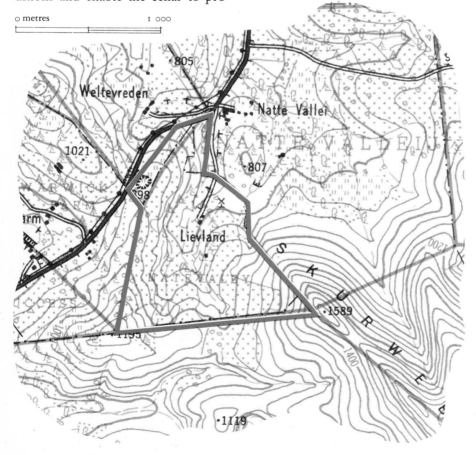

duce a broad selection of wines.

Lievland was originally part of the grazing ground alongside Natte Vallei, the Klapmuts farm (granted in 1715 to Jurgen Hanekom, a Dutch East India Company wood-cutter) that was incorporated with Abraham de Villiers's historic property in 1819. During the following year, De Villiers sold part of this extra land (now the Le Bonheur wine Estate) to his brother Jacob, and a further portion to Daniel Moller, who built a cellar and a gabled house on land that he named Beyerskloof — today known as Lievland.

This property passed through the hands of a succession of owners before it was inspected and bought by the Baron and Baroness Von Stiernhjelm in 1934. Originally resident in the southern part of the Baltic state of Estonia (now part of the Soviet Union) called Livonia, the Von Stiernhjelms felt the futures of their young children threatened by a belligerent Germany and an acquisitive Soviet Union, and looked forward to a new life in South Africa. The Baron died before the family left Europe, and the widowed Baroness brought her young child to a new home at Beyerskloof in the Simonsberg hills in 1936. Almost penniless and totally inexperienced in farming and wine production, Hendrika

All of Lievland's vineyards are planted on sloping ground on the sides of a basin-like valley.

von Stiernhjelm used the assistance of neighbours and her own indomitable spirit to support her family and create a quality wine-making tradition on the small farm she renamed Lievland, after the birthplace of her husband and their children. Mrs. Von Stiernhjelm's wines and farm produce were sold, often from door to door, throughout the Boland for over twenty years.

In 1964 Mrs. Von Stiernhjelm sold Lievland to Hans Steenkamp, a sheep farmer from the Karoo, who took no interest in the farm cellar and joined a local co-operative. For nine years Hans harvested grapes from the vines planted by Mrs. Von Stiernhjelm, delivering them each season to the co-operative cellar. In 1973 the farm was sold to Gert van der Merwe and Dan Benade. Gert was born and raised on a fruit and potato farm in the Ceres–Karoo and had operated an independent packing house for his own export fruit. Though he had had no experience of wine making, Gert saw a possibility eventually to produce Estate wine in the farm cellar. He therefore refused the opportunity to join the co-operative, choosing instead to send his grapes to a private cellar.

Gert found the old vineyards to be severely run down and planted in insufficiently prepared ground. As a fruit and potato farmer, he understood the value of the thorough preparation of the soil for the vine roots, and he began to replant and develop his relatively small farm. Lievland has 70 hectares of vineyards with the proportion of white to red balanced in favour of white. In 1981 Gert built a relatively small cellar behind Mrs. Von Stiernhjelm's cellar and equipped it with all the facilities needed to make quality wines. All the farm's grapes were sold to a Paarl cellar until 1982, when Gert and Janey Muller made the first wine at Lievland since 1964. He had met Janey, a graduate of Stellenbosch University's wine-making course, in that same year, and persuaded her to join the team in an advisory role — to make the major decisions and to teach Gert the procedures involved in converting healthy grapes into quality wine.

Faced with the unknown difficulties of marketing a new range of wines, Gert reduced the size of the problem by deciding to continue to sell a proportion of the crop as grapes to the Paarl cellar.

The warmer and cooler slopes around the basin provide grapes with varying degrees of sweetness. During most seasons, some grapes have to be picked early to maintain sufficient acidity, while others may be left to make more use of the exposure to sun. In a good ripening year, like 1982, all the vineyards are able to ripen their grapes fully without acidity loss. The harvesting team bring Chenin blanc, Bukettraube and Cape Riesling grapes to the cellar at mid-ripe and full-ripe stages for the production of dry and semi-sweet wines. The grapes are crushed and the skins allowed to stay mixed in the juice for several hours to obtain extra flavour. The skins are pressed and the composite free-run and pressed juice is inoculated with yeast and allowed to ferment at cool temperatures (12-15 °C) until dry, or, in the wine blended from Chenin Blanc and Bukettraube, until the required degree of sweetness has been reached. The fermentation is normally arrested by lowering the temperature of the must and raising the SO_2 level, retaining the natural sweetness.

Lievland also made the first red wines in its cellar for a generation by mixing, at the crusher, the produce of their young Cabernet Sauvignon vineyard with a part of their Cinsaut. This wine has its initial fermentation stage in the Estate drainer, where the liquid must can be pumped over the skins to obtain the required colour. The result is a moderately coloured, fruity must that has been fermented out in closed containers and then placed in oak casks for maturation. Gert intends to continue producing this Cabernet-based red wine, and plans to add the produce of his Merlot vineyard, planted on the cool, south-facing slopes, to the blend. A pure varietal wine from Cabernet Sauvignon is also expected to be added to the range.

Wine is for sale on the Estate.

Meerlust

Grapes that ripen slowly, through an extended season of moderate temperatures, are believed to make the best wines. The required average temperatures occur in regions far from the equator, on elevated slopes and plateaux and in areas where prevailing winds reduce temperatures and the effect of the summer sun. The Cape's wine-growing areas nearest the ocean, like Faure, south of Stellenbosch, where the Meerlust Estate is found, benefit from the humidity and reduced temperatures brought by on-shore winds that blow during the summer. The humidity and the cooling effect of the air often bring to the vineyards summer evening mists that remain until they are evaporated by the heat of the day.

There is only one farm between Meerlust and the sea, and Meerlust's humid conditions are similar to those enjoyed by some of the great Châteaux of Bordeaux. Though Meerlust is close to sand dunes, there is little sandy soil on the Estate, and the two main soil types found in the vineyards are moderately fertile. The white cultivars are grown on the alluvial, deep-draining soils by the Eerste River. During harvesting, the grapes from these vines are sold to a wine merchant for use in the production of blended wines. The red varieties are found mostly on the pebbly loam soil that owes its origin to the Helderberg Mountains behind the farm, and the wine from these varieties is made and matured on the Estate.

Nicolaas Myburgh, owner of Meerlust, is a meticulous farmer. He believes that a wine Estate can make its best wine by choosing shy-bearing cultivars, pruning them to gain a relatively high yield and irrigating them to achieve this production without placing undue strain on the plant. To do this, however, the Estate must have sufficient water to cope with the possibility of a dry winter, as happened in four of the first five years of the 1970s in the Stellenbosch area.

Henning Hüsing, the first owner of Meerlust, is believed to have been South Africa's first millionaire. He became the wealthiest citizen of the fledgling colony at the Cape, and op-

Eight generations of Myburghs have owned Meerlust, a direct line of ownership extending from 1756 until the present day.

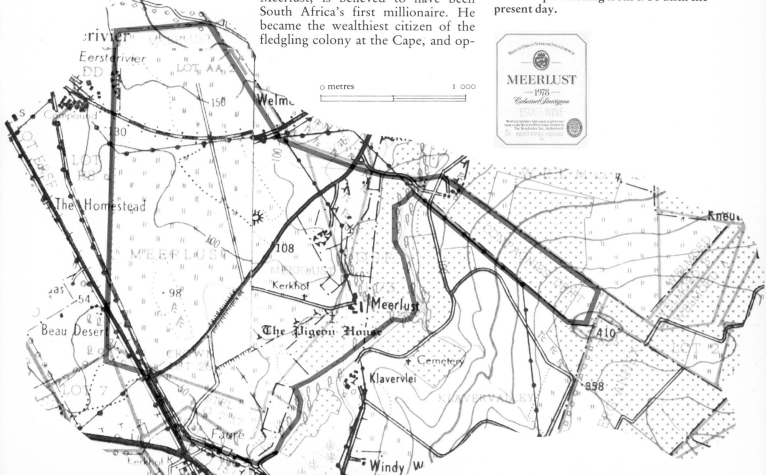

erated numerous enterprises and held many public offices from his manorial home by the Eerste River at Faure. In 1672, when Hüsing arrived at the Cape in the *Azia*, he was described as a shepherd from Hamburg. Willem Adriaan van der Stel later said that Hüsing was very poor during the early years, and claimed to have helped establish him in business.

To provide meat for the inhabitants of the Colony, settlers and Company employees were allowed to use land in the interior to graze cattle and sheep after 1673, and Hüsing became involved in the production of meat. In 1678 Governor Johann Bax provided land in the area now known as Somerset West for five farmers to graze animals and grow grain. Hüsing was one of the fortunate group and was declared a free burgher during the same year. In 1680 he established his base on the Eerste River on the farm he was later to call Welmoed, and by the time of his marriage to Maria Lindenhovious (a maid from the Van der Stel household) in 1704 was the owner of great flocks of cattle and sheep. During the following year Simon van der Stel awarded Hüsing a contract for the supply of meat to the Colony's hospital, navy and army garrison for fifteen years. In 1685 he was appointed a Heemraad and was the most important official in the Stellenbosch area after the Landdrost.

Soon Hüsing became a major landowner, obtaining freehold ownership of Welmoed in 1690 and an adjoining property, which he named Meerlust, in 1693. He built a grand house on Meerlust and sold Welmoed to his friend Jacobus van der Heyden in 1698. He had 100 000 vines in the Meerlust vineyards and was described as the most important wine farmer in the Cape, with wine production outranking that of ex-Governor Simon van der Stel on his Constantia Estate.

His fortunes began to change during the early 1700s as Governor Willem Adriaan van der Stel, envious of the wealth being amassed by Hüsing and his friends, used his official capacity to develop his own Somerset West property and accumulate his

Though Meerlust is near False Bay and benefits from high humidity, irrigation from dams is provided for vineyards.

own fortune. In 1704 Hüsing reported Van der Stel's activities to the Directors of the Company in Holland, but no action was taken. Shortly afterward the Governor rescinded the meat contract and Hüsing and his associates prepared a petition reporting the position to the Company officials in Europe. Van der Stel acted promptly, arresting the five leaders of the revolt in 1706 and shipping them to Holland in disgrace. Unknown to Van der Stel, the revolutionaries smuggled with them a copy of a document written by Adam Tas, explaining the position at the Cape. The Company Directors ordered an investigation and subsequently the Governor was removed from his post. Hüsing was able to return to his beloved Meerlust in 1708. He died in 1713 and his widow sold Meerlust to a wealthy Cape Town builder, Blanckenberg, in 1716.

Johannes Albertus Myburgh bought Meerlust from the Blanckenberg family in 1756 and used the property to graze cattle and sheep, growing a few vines for the production of fortified sweet wine. Since Johannes, eight generations of Myburghs have owned and farmed Meerlust, one of the longest lines of direct descent in an area noted for successive generations of farm ownership. One of these Meerlust Myburghs, Philip, who lived on the farm around 1800, was a noted explorer. He was the first to follow the course of the Orange River, and lived for many years on the eastern frontier of the colony, during which time he sketched out a plan for the development of the district of Graaff-Reinet.

Lady Anne Barnard was entertained sumptuously in the home of Mr. and Mrs. Philip Myburgh at

Each of the buildings in the Meerlust complex has been renovated by Nicolaas Myburgh, the present owner.

Meerlust at the end of the eighteenth century, and a few years later, following the Battle of Blouberg in 1806, General Janssens took refuge with the Myburghs after the British had won the Cape for the second time.

The present owner and vintner of Meerlust Estate, Nicolaas Myburgh, took over the farm from his father in 1950. At that time the only wine varieties on the farm were Steen and Green Grape. The vineyards covered about 85 hectares and no serious attempts had been made to produce high-quality wine. Meerlust has changed appreciably under Nicolaas Myburgh's control. There are now over 250 hectares under vine, mostly planted with red wine varieties.

"I believe that because of the moisture the sea breezes provide," says Nicolaas, "we're capable of producing really great red wines, of international stature. We're doing so now, and are only just starting to realise it. We receive what could be the highest amount of dew in the world, about 100 mm per year, being so close to False Bay. This dew can often still be seen on the leaves of the vines as late as eleven o'clock in the morning. But when you add this to our 650-mm rainfall on Meerlust, the total is still less than the vines need for the best results, as we have a very high rate of evaporation. So I use spray irrigation, and by calculation I try to add about an extra 250 mm during spring and early summer. If we have a wet winter, as in 1974, I give the vines less; in a dry year, more."

The Meerlust Estate specialises in the production of red wine. It has 200 hectares of red variety vineyards and only red wines are produced in the Estate cellar. The juice of Cabernet Sauvignon, Merlot, Cabernet franc, Pinotage and Pinot noir grapes are fermented and the wines made are matured and made ready for bottling in the Estate cellar. The grapes produced by the Cinsaut vines are delivered to a nearby wine co-operative for crushing and fermentation.

Five hectares of Sauvignon blanc vines fill a gap in the red variety harvesting season, and the white grapes are delivered to a Stellenbosch wine merchant's cellar.

1. PINOT NOIR
2. CABERNET SAUVIGNON
3. STEEN
4. SAUVIGNON BLANC
5. HANEPOOT
6. MERLOT
7. PINOTAGE
8. CABERNET FRANC

Meerlust 157

Built in the late 1700s, possibly by a Myburgh, Meerlust's dovecote is one of the country's two oldest.

Nicolaas Myburgh (right) and his manager Japie Vermeulen assess the new vintage.

By planting Merlot in 1973, and extending the vineyard during each subsequent season, Nicolaas Myburgh became one of the first producers with a commercial quantity of wines made from the traditional Bordeaux blending varieties. The wines made from Meerlust's extensive Cabernet Sauvignon vineyards were first released in pure varietal form in 1978 and a portion blended with Cabernet franc and Merlot was first marketed under the name Rubicon at the end of 1982.

The grapes were picked at 22-23 ° Balling and fermented in both open concrete and closed tanks at around 25 °C. Giorgio Dalla Cia, Meerlust's cellar master, retains the skins within the fermenting must until the sugar has reached 3 ° Balling, and sometimes an even lower point, before removing the liquid to ferment dry. The wines are racked once and encouraged to start malolactic fermentation, after which they are filtered clean, ready for maturation. All the wines made from the produce of the premium red varieties are matured in the Meerlust oak maturation cellar, which has both Yugoslavian 4 000 litre and French small oak casks, to be able to provide up to eighteen months of wood contact while ageing. As more of the young vineyards begin to bear to their

Spray irrigation is systematically used to add the equivalent of 250 mm of rainfall.

full capacity, this cellar will be extended to cope with the extra production. All the red grapes are crushed separately and the wines are fermented and matured individually. When the Cabernet Sauvignon, Cabernet Franc and Merlot wines are considered to have gained sufficiently from their period in oak, they are blended in the proportions chosen by Nicolaas and Giorgio at that time.

"When we have a very wild and powerful Cabernet Sauvignon, we add more Merlot to the blend," says Giorgio. "When the Cabernet is less aggressive, we will increase the quantity of the two Cabernets in the blend." The Cabernet franc variety produces wine with less of the untamed character that requires such demanding maturation of pure Cabernet Sauvignon, and reaches maturity at a much earlier stage. The Merlot has a natural affinity with the Cabernet varieties, having a similar 'grassiness' in both aroma and flavour when young.

It has a sweeter, softer taste than both, with a very fragrant nose, and is ready to drink at an earlier age than Cabernet Sauvignon.

Once blended, these wines are put in oak for a final marrying phase before they are transported to the cellars of a Stellenbosch wine merchant for bottling. Part of Meerlust's heroic quantity of Cabernet Sauvignon is bottled separately as a pure cultivar wine to satisfy the demand that was created by the release of several vintages before the blend became available.

Meerlust Pinot Noir is made along roughly the same lines as the Cabernets and the Merlot wines, but receives a shorter wood maturation period and is bottled as a pure cultivar wine.

Nicolaas and Giorgio have been joined on the Estate by Nicolaas's son, Hannes, latest in the long line of succession. Hannes, trained in cellar technology, is assisting Giorgio, learning the handling and care of the red wines produced by Meerlust's celebrated vineyards.

Wine is not for sale on the Estate.

Middelvlei

Most of the choice pieces of level farming land around Stellenbosch had been granted to pioneer farmers by 1720. These were normally small pieces of agricultural ground and were expanded by perpetual quitrent grants handed out during the period after 1813. But large tracts of land on steep slopes on the sides of the valley were still Crown Land in 1900 and were transferred to the Municipality of Stellenbosch soon after that time. Middelvlei was part of a grant of about 850 hectares of land around the Papegaaiberg, a forested hill to the west of Stellenbosch, made to the Municipality in 1908.

In 1923 two sons of Jan Hendrik Momberg, the owner of the Vlaeberg farm to the south-west of Stellenbosch, bought less than two hectares of this land on the western slopes of the Papegaaiberg to build houses and a small cellar. At the same time the brothers, Tinnie and Niels, leased adjoining land from the local authority for the planting of vineyards. The situation remained thus until they bought 100 hectares of the leased land from the Municipality in 1936. Two years later the final piece of land was added, extending Middelvlei to its 160 hectares. The farm is adjacent to the town of Stellenbosch in a horseshoe-shaped basin with gentle slopes facing east, south and west toward the central depression where the Estate's house and cellar are located.

Tinnie and Niels Momberg planted orchards and vineyards on their property and made fortified dessert wines and rebate wine for brandy production. They raised their families on the land they shared, and each named his eldest son Jan Hendrik Momberg in honour of the grandfather. When Tinnie died in 1942 he left his share of the farm in trust to his son Jan who was only a few years old. The property was farmed by Niels until his death in 1960, when the two cousins with the same name took over and operated Middelvlei as a partnership, as their fathers had done. To distinguish them from each other, the Stellenbosch people began to call Niels's son, who has a lot to say, Jan Bek, and his quieter cousin Stil Jan. In 1963 Jan Bek discovered that the manorial Neethlingshof Estate was for sale and took the opportunity to buy one of Stellenbosch's top vineyards together with its magnificent house. His cousin then became the sole proprietor of Middelvlei.

Traditionally, the Middelvlei vineyards have contained more red vines than white. In 1960 the proportion was in excess of 80 to 20 and a substantial quantity of Cinsaut grapes was crushed in the farm cellar each year. Many of these vines had been planted by Tinnie and Niels Momberg and were replaced by Jan in the

The high standard reached by Jan Momberg's Pinotage is now being matched by the Estate's Cabernet.

Cinsaut fell out of favour during the late '70s. Middelvlei has only a small area left.

Colour is quickly extracted from Pinotage skins as the fermenting must is pumped over the husks.

On the edge of Devon Valley, the Middelvlei vineyards slope to the south, east and north.

late '60s, mainly with Chenin blanc and Pinotage. To cope with the production of his new white vineyard and to supply the developing market for white table wines, Jan rebuilt the old cellar and installed cooling equipment.

The Middelvlei Estate, located on decomposed granite slopes alongside Devon Valley, has mostly Hutton soils, containing a percentage of clay, that retain sufficient moisture from winter to ripen the crop of grapes in summer. This type of soil, even on Middelvlei's cooler slopes, has proved to be well suited to the growing of red varieties and the production of medium and full-bodied red wines. Jan planted Cabernet Sauvignon and Tinta Barocca in the 1970s to supplement the Pinotage and Cinsaut vineyards. During this period, the Steen and Clairette blanche vineyards were also extended and blocks of Chardonnay and Sauvignon blanc were planted more recently, bringing the total of white variety vines slightly in excess of the red. All the vineyards, with the exception of Cinsaut, are trellised and all are grown without irrigation.

Jan Momberg's first Estate wines were bottled from the 1972 harvest of Pinotage. His early wines were full-bodied and heavy, made from late-picked grapes. After several vintages he changed styles, harvesting the Pinotage earlier and aiming for a lighter, more fruity type of wine.

The grapes are picked mid-ripe (22-23 ° Balling), and are fermented in the mash. Jan removes the skins once satisfactory colour has been extracted and presses the skins hard to obtain all the available liquid. All of this pressed must is returned to the fermentation tank to ferment out the rest of the sugar. After fermentation is complete, the wine is stored in closed concrete tanks, where it undergoes malolactic fermentation. In about October of the vintage year, Middelvlei's red wines are transferred to a Stellenbosch wine merchant's cellar where they are matured in large oak casks, and bottled about one year later.

In addition to the consistent appearance of successive vintages of Middelvlei Pinotage on the South African wine market, Cabernet Sauvignon from this Estate will be released each year, starting with the 1981 vintage. The wine is made in a similar style to the Pinotage, but receives a slightly longer period of maturation in wood.

Although more than half of the Estate's vines belong to white grape varieties, it is unlikely that there will be a white wine bearing the Middelvlei label within the foreseeable future.

Wine is not for sale on the Estate.

Montagne

Facing the south-easter as it comes through the gap over Devon Valley, the Estate is often bathed in morning mist.

Though this was for centuries a mixed farm, the granite, gravel and shale soils have proved to be ideal for vineyards.

The potential of this, one of the Cape's most important wine Estates, was only realised in the mid-1960s, though vines had been grown on the farm without pause since 1692. Until the establishment of a block of Cabernet Sauvignon vines by the father and son team of Maurice and Walter Finlayson, the property had chiefly been used to graze cattle and sheep, though many types of mixed farming had been undertaken. But it was when the Finlayson's hand-made Cabernets of 1968 and 1969 were offered to wine lovers that the full scope of the farm's promise was glimpsed.

This property was originally called Het Hartenberg by two young bachelors, Christoffel Esterhuyzen (or Estreux, as his name was spelt before he arrived) and Coenraad Boom, who settled on the farm in 1692 soon after arriving in the Cape. During their first year, the two planted 2 000 vines and reaped 20 bags of wheat from the three they had sown. They had 200 sheep and 12 oxen and this may be considered a pretty good start.

In 1704 Christoffel Esterhuyzen was given the official title to Het Hartenberg, and ten years later he bought the adjoining farm Weltevrede. He married Elizabeth Beyers, daughter of the Company wagon maker, and in 1710 they had a son, Christoffel. It appears that they lived on Weltevrede, and that Hartenberg was without a substantial house throughout the next century.

In 1721 Esterhuyzen sold Hartenberg to Paulus Keyser, a young German who had emigrated as a soldier in the service of the Company in 1707 and worked as a farm labourer until declared a free burgher in 1715. Keyser married Aletta Lubbe, daughter of Barend Lubbe, owner of Neethlingshof, and was known as a leading hunting guide, participating in many elephant hunts.

Jacob van Bochem bought Hartenberg in 1725 and added Weltevrede to it again the following year. He had been the accountant in the Company butchery, and later was granted the liquor retail monopoly. After passing through the hands of Ari Lekkerwyn, the combined farms became the property of Aron van Ceylon, a slave who had served his time and had been given burgher rights. Somehow both farms reverted to Keyser at some time during the following nine years, as it is recorded that his son-in-law bought them from him in 1745.

In 1838 the brothers Jacobus and Johannes Bosman, whose family ran a number of farms in the Bottelary area, bought the twin properties, and in 1849 Johannes bought his brother's interest and built the manor house that overlooks the north-south valley running through the centre of the farm. Like the previous owners, the Bosmans grew grain and grazed animals on the long gentle slopes of Hartenberg, with vines grown only for domestic use. The vineyards were extended by the Hauf family after they bought Hartenberg in 1928. Mrs. Hauf also planted many trees around the house and cellar, thus greatly enhancing the beauty of the property.

In 1948 Hartenberg was bought by Dr. Maurice Finlayson and his wife. The doctor brought his young family to live on the farm, but continued to practise medicine in Cape Town. The farm had been bought to provide him

The original house has been restored and visitors may once again buy wines on the Estate.

To re-establish Montagne as one of the nation's leading wine Estates, Gilbey's have extended both vineyards and cellar facilities.

with a retreat from city life and a chance to indulge his interest in farming. Faced with 180 hectares of mostly open space, he was able to indulge in a number of hobbies. A dairy was begun, fruit trees and vineyards were planted and poultry was reared for the production of eggs. Maurice Finlayson had four sons, and as the boys grew up they became interested in the various projects initiated by their energetic father.

When he was a schoolboy, Walter, Maurice's eldest son, was attracted to the idea of dairy farming, and involved himself deeply in the farm's dairy activities. But his father began to turn his attention increasingly to wine making. Though he was by profession a pathologist, his greatest interest lay in the field of microbiology, the study of minuscule living organisms. In the production of wine, knowledge of the exact nature of the activity of yeast and bacteria is of immeasurable importance, and Maurice's ambition for Hartenberg became to make it a wine farm of international standard.

At the time of Maurice's purchase Hartenberg's grapes were sold each year to a wholesaler, and the farm cellar had not been used for many years. After Walter left school, he spent two years in Europe studying agriculture, and on return to the farm found the old cellar making wines of indifferent quality. In collaboration with his father, Walter developed plans for improving the cellar and extending the vineyards. In 1960 he was placed in charge of the vineyards and cellar and made his first wines.

During the '60s Walter made red wines from Cabernet Sauvignon and Cinsaut and sold them to a wine merchant at good prices. They were then blended into the now-discontinued range of Vlottenheimer wines. Without the facilities of cold fermentation, and using as stock the robust white varieties present on the farm, Walter made white wine only for the production of sherries.

In 1968 Walter bottled 250 cases of Cabernet Sauvignon, and sold them all in a matter of weeks. In 1969, 1000 cases went in about the same time. This indicated that a new market was opening up. After these pioneering successes, Walter and his father toured South Africa with samples of their Estate red wine in an effort to see just what could be expected if they were to embark on wine making on a large scale. The results were far more spectacular than they had ever dreamed of, and they returned with orders that they could not fulfil.

The first year that the farm was able to offer the public a substantial quantity of high-quality red and dry white wines was 1971, when Walter produced his first 'Superior' classification wine, Montagne Cabernet Sauvignon 1971. The name Montagne was chosen for the Estate wine products because Hartenberg had already been registered by a wholesaler and was unavailable to the Finlayson family.

Walter decided to spend a great deal of time and money in modernising the cellar, for he had also found that there was a far higher demand for good-quality dry white table wine than he had expected. The year of change was 1973, when extensive replantings of white varieties such as Cape Riesling and Chenin blanc, as well as further planting of Cabernet Sauvignon, Shiraz, Pinotage and Cinsaut, were undertaken.

Montagne is a long narrow Estate, running down from the granite outcrop of the lower Bottelary Hills to the north, along both sides of a shallow valley, to level sandy and shale soils, covering a distance of 1,3 kilometres and providing a difference of almost 150 metres between the highest and lowest vineyards. The soil on the higher slopes has been produced by decomposed granite. In the centre of the farm, gravel soils predominate,

The Estate vineyards lie on slopes facing north and south, on either side of a damp, shallow valley.

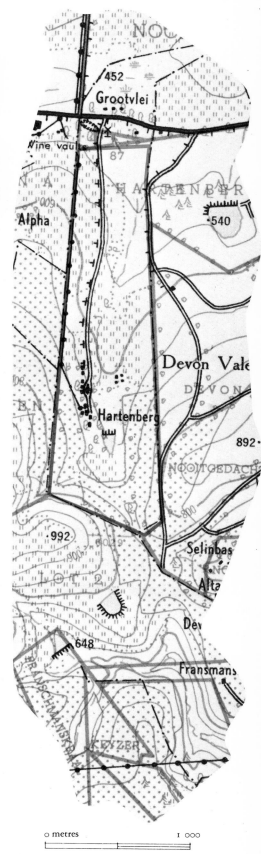

and an earlier owner sold some of this valuable vineyard soil for road making. The lower vineyards are planted in soil derived from shale. All three types have clay content or are firmly based on a layer of clay, providing the farm with a moist water table for the roots of vines or fruit trees. The centre of the valley, running down the spine of the farm, acts as a drainage furrow for seepage from the mountain and the slopes of the farm, which face south-west and north-east into the valley. With the choice of soil types and warmer or cooler slopes, and with moisture-retaining soils, Montagne is capable of providing ideal conditions for both early- and late-ripening cultivars.

Walter Finlayson planted Cabernet Sauvignon, Shiraz, Pinotage and Cinsaut vineyards on the red, granitic soils, and Zinfandel on the lighter, gravel soils. He planted Chenin blanc on the red ground and Cape Riesling on the shale. Each vineyard produced wines of notable quality and Montagne became a leading producer of Estate wines during the early '70s. Walter was joined by his brother Peter for the 1976 harvest, and later that year Montagne became the property of Gilbey's, who have maintained the schedule of replanting with top varieties.

The growing conditions on Montagne suit the production of light- to medium-bodied fruity wines. The average rainfall of 675 mm is above average for the area and the farm is exposed to the south-easter as it blows in from the sea on most summer days. Montagne has rows of pines, grown as windbreaks to reduce damage from the prevailing summer wind, and these allow the vines to benefit from the ventilation without harmful effects. About 60 per cent of the vineyards receive irrigation in summer.

Though the grapes of most varieties are able to reach high sugars, experience has shown that the farm produces best when the grapes are picked mid-ripe (21-22 ° Balling). The Cabernet Sauvignon is traditionally left to reach 22-23 ° Balling and then fermented at warm temperatures (average 26 °C) to gain extra colour. The mash is fermented down to approximately 6 ° Balling to gain extra depth of flavour, before the skins are removed. Though the Shiraz grapes are normally picked with slightly less sugar, a similar procedure is followed through fermentation.

The practice is changed for the production of Zinfandel, with grapes picked earlier (21-22 ° Balling), the must fermented at a lower temperature and the skins removed from the mash when half the sugar has been converted. The Zinfandel is fermented dry in closed steel tanks and is not aged in wood. The Cabernet Sauvignon and Shiraz wines receive at least one year in wood.

From the Estate's white vineyards Sémillon and Chenin blanc are harvested full-ripe to make broad, full-bodied wines, and Cape Riesling is picked early to provide a natural balance. The white wines are fermented in steel at cold temperatures (13-14 °C) until dry, and are cleaned and bottled, ready for bottle maturation.

Wine is for sale on the Estate.

Montagne 163

Muratie

The history of this famous Estate is shrouded in the mists of time. It is known that it was one of the first mountain farms to be settled in South Africa. Most of the early Stellenbosch farms were situated on the flat river land to the south. Muratie is in one of the beautiful valleys that run down from the Simonsberg toward the town of Stellenbosch.

Lorenz Campher, a German who must have immigrated earlier than the French Huguenots, chose the lower valley slopes for his home around the year 1685. We know that he was still there fourteen years later, for he was granted freehold title to twenty-five hectares of the valley in 1699 by Willem Adriaan van der Stel.

Virtually nothing is known of the farm during the next century, though it is unlikely that the Campher family remained there for much of that time. It is known that Martin Melck's wife bought the property about 1770, at which time the old home was in ruins, probably as a result of fire.

Shortly after the purchase the farm was given to Melck's son-in-law, Jan David Beyers, a builder by profession, who rebuilt the house, renaming the farm Muratie, a corruption of the word *murasie* (ruin). Three Beyers generations lived on Muratie. Jan David's grandson extended the property to well over 85 hectares.

It is likely that the Beyers family made wine in considerable quantity, for many old vines were found on the farm by the owners who followed. The cellar also dates from the time of the Beyers family. As an oak next to the cellar is believed to be over 200 years old, it is possible that Beyers rebuilt the cellar at the same time as he worked on restoring and extending the house. Parts of the present cellar may well be also over 200 years old.

Georg Paul Canitz came to South West Africa in 1907, virtually by accident. Under treatment for an illness, he was advised to take a sea voyage, and during the trip his health deteriorated to the point that he had to be taken ashore at Swakopmund. After his recovery, he decided to remain in Southern Africa and was soon joined by his wife and family. They moved to South Africa in 1910, living first in Cape Town and later in Stellenbosch.

Canitz was a talented artist and supported his family in this manner for several years. Then came the day when he swapped his palette and brushes for a wine cellar, although before this time his knowledge of the wine industry had been limited to an appreciation of the end product.

Canitz first saw Muratie from horseback, just as Campher must have done. He was riding in the Simonsberg area with his family, was attracted to the farm, and mentioned his discovery to his close friend, Professor Perold of Stellenbosch University, head of the University's agricultural and viticultural faculty. Perold immediately advised him to buy the farm. He considered that it was one of the best red wine farms in South Africa and that it would have great prospects, properly farmed.

Georg Paul Canitz bought Muratie in 1925. As Canitz knew little about wine making Professor Perold spent a considerable part of his spare time over the next two years on the farm, selecting from the Riesling, Cabernet Sauvignon and Cinsaut vines, and getting the old cellar into full working

Muratie is a corruption of a Dutch word meaning ruin. This home is believed to have been rebuilt after a fire.

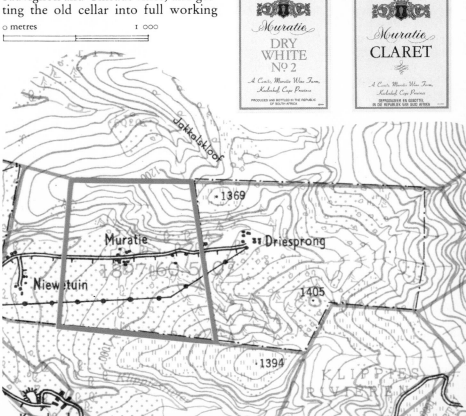

(upper) Ben Prins, vintner, tops up wine lost by evaporation. (lower) The rustic finish of several of the Muratie buildings has seen little change in two centuries.

Muratie was the only Estate making wine from Pinot noir grapes from the 1920s until the mid 1970s.

Muratie mountainside vineyards slope to the west and north, well suited to the ripening of red cultivars.

order again. In the desire to add a little extra to the South African wine industry, Professor Perold obtained several canes of a recently imported Pinot noir clone for his friend to try on his new farm. The experiment proved a great success.

Professor Perold trained Wynand Viljoen as Canitz's foreman, teaching him the finer points of good wine making. Later Viljoen taught Ben Prins the skills of maintaining a vineyard and running a cellar, and in time Prins was to take Viljoen's place as foreman and cellar master on Muratie. When Georg Paul Canitz died in 1959, his daughter Annemarie took over management of the farm.

Miss Canitz has continued with the varietals that Professor Perold established on the farm. Her dry fermented wines have been limited to Cabernet Sauvignon, Riesling, Cinsaut, Steen and Pinot Noir. The Canitz family have been making each of these fine dry wines for sale to the public as Estate wines since the late 1920s.

Muratie has 110 hectares of which 65 are vineyards. As the pride of the Estate is its red wines, Cabernet Sauvignon and Pinot Noir, there are many more red vines than white. During the late 1960s most of the older vineyards had to be replaced, and Ben Prins extensively replanted. Prins has not switched to new varieties, or moved the emphasis of the Estate from one type of wine to another, so Muratie's future wines will be essentially the same as those made by Georg Paul Canitz, Wynand Viljoen and Annemarie Canitz.

The wines on Muratie are made by spontaneous fermentation. This is a natural process, in which the grape skin, when covered by 'bloom', has enough yeast to begin fermentation as soon as the skin is broken and the 'bloom' comes in contact with the sugar in the juice.

Ben Prins makes a very popular port-type wine from a mixed vineyard of port cultivars. Seven varieties produce grapes that attain high concentrations of sugar on untrellised bunches. The grapes are crushed together, allowed to ferment for a short period, and then fortified with brandy. Most of this wine is sold within a year of harvesting. A small quantity is aged for five years before sale.

The equipment in the Muratie cellar is vintage material. The red wine concrete tanks being used today were installed during the 1930s on the recommendation of Roberto Moni, one of South Africa's best-known wine merchants. The Estate's wine is sold on the farm. This occurs at irregular intervals, for the wines are bottled at different times, with Muratie Cabernet Sauvignon spending up to eighteen months in wooden casks. Wine lovers wishing to buy some of Miss Canitz's remarkable wine are advised to place orders for the wines wanted.

Neethlingshof

Though the house was built in 1814, both farm and home were renamed Neethlingshof in 1844.

Jan Momberg, champion of greater individuality in Cape white wines.

Jannie Momberg grew up on his family's property, Middelvlei, adjacent to the town of Stellenbosch. As a boy, he wandered through the vineyards and orchards of Devon Valley and across foothills of the eroded Bottelary Hills marvelling at the beauty of the setting and the gracious homes of the Cape's rural aristocracy. The closest of the manorial Estates to Middelvlei was Neethlingshof, with almost 300 hectares of east-sloping land and an elegant gabled house facing the Simonsberg and Stellenbosch.

In 1960 Jannie inherited his father's share of the Middelvlei partnership. For three years he continued to operate the farm in partnership with his cousin, also Jan Momberg, but when in 1963 Jannie heard that Neethlingshof was for sale, he sold his share to his cousin and moved into the big house with the great avenue of stone pines.

Neethlingshof was originally known as Wolwedans and was settled by Barent Lubbe, a German immigrant, soon after his arrival in the Cape in 1690. Lubbe worked on the newly developed farms in the vicinity of Stellenbosch for several years, and is recorded as being a member of a work team instructed by the Governor to do maintenance work on a drainage channel near the church building in 1692. In the same year he occupied the land he called Wolwedans, and in the census for that year he is recorded as living alone on his land with a hundred sheep, eleven cattle and a pig. He married Jacoba Brandenburg, who was a niece of a junior merchant in the service of the Company at Cape Town, and was granted Wolwedans in 1699 by governor W. A. van der Stel. In 1715 the Stellenbosch church council gave him a liberal gift of 40 guilders, something they only did in extreme cases, and one can presume that Lubbe was finding the life of a pioneer in Stellenbosch's timber and scrubland difficult. By 1716 he was living at Wolwedans without a wife, but with eight children and a modest list of possessions.

In 1717 the farm was sold to Joost Hendriks van Rheenen, who was apparently unmarried. Van Rheenen sold Wolwedans in 1724, and the farm passed rapidly through many hands. One of the owners during this period was Jan de Villiers, son of Pierre de Villiers, who inherited Laborie in 1753 and bought Wolwedans in 1761. In 1767 he and his wife, Anna Hugo, were murdered by a slave on the Vredenburg farm, beside the Eerste River, just below Wolwedans. Both Laborie and Wolwedans were inherited by Jan's younger brother Isaac, and both were sold out of the family by Isaac's widow. In 1788 the farm was bought by Charles Marais, great-grandson of one of the most capable Huguenot farmers, who bore the same name.

Charles Marais extended the vineyards and turned the Estate into a successful farming operation during a period of prosperity in the Cape wine industry. He built the manor house in 1814, and when he died in 1816 his widow and children continued to run the farm, which was extended by a 200-hectare grant in the same year. In 1825 they were recorded as having 80 000 vines on Wolwedans and nearby Fransmanskraal, and 30 leaguers of wine in the farm cellar. When Magdalena Marais died in 1844, her son-in-law Johannes Neethling became the owner and changed the name of the property to Neethlingshof. In 1870 Neethling sold the farm to Jacobus Louw, and Neethlingshof remained in Louw hands for almost a hundred years. During this period the farm of almost 400 hectares was divided, with the smaller portion reviving the name Wolwedans. The larger section, containing the Marais–Neethling homestead, was bought by Jannie Momberg from the Louw family.

When Jannie moved to Neethlingshof, Gys van der Westhuizen was established as the farm manager, and together they formed a management team that has expanded the vineyards to provide a large crop of grapes for the modern cellar that Jannie built as an extension of the old Marais winery. Running from high on the slopes of the Bottelary Hills down towards the valley of the Eerste River, Neethlingshof has decomposed granite soils, of the Hutton type, and sandy loam soils. Parts of the farm are damp and are unsuitable for vines. The majority of the vineyards are trellised and only the Cinsaut vineyards are grown as bush vines. Thirty per cent of the vineyards are irrigated. Facing east, on fairly easy slopes, the vineyards have reduced summer afternoon exposure to the sun, and are provided with a degree of temperature moderation and ventilation by the gentle effects of the prevailing southeast wind.

When Jannie came to Neethlingshof he found that many vineyards contained Chenin blanc vines, and these have been retained. He has planted Colombard, Clairette blanche, Weisser Riesling, Gewürztraminer, Kerner and Bukettraube, and now the proportion of white vineyards to red is almost two to one. The Cinsaut vineyards he found have mostly been replaced by Pinotage, and a small vineyard of Cabernet Sauvignon was planted in the early '70s.

Schalk, Gys van der Westhuizen's son, worked around the cellar and in

The avenue of stone pines leading to the Neethlingshof cellar covers over a thousand metres.

Colombard grapes at Neethlingshof produce moderately fruity wines with sufficient body and acidity to mature well.

the vineyards during his childhood on the farm, and went on to study cellar technology at Elsenburg Agricultural College. After graduation he became Neethlingshof's cellar master. In consultation with Jannie, the father and son team carry out the Estate's development programme, which is aimed to produce full-bodied, dry white wines with deep flavour, and light- to medium-bodied reds with the ability to mature early.

"I believe that we have the best white wine techniques in the world," says Jannie. "With cold fermentation, we can duplicate the German and French temperature conditions for pressing and fermenting, and we have the advantage of being able to choose our moment of ripeness. They have to take a chance with their climate for ripening, but their fermentation temperatures are perfect. Before we had cold fermentation methods, we had to learn to make the best of what we had. When you combine that skill with the use of constant cool cellar temperatures and the advantage of guaranteed perfect ripeness, then you know you can make great wines each year. In France or Germany they have

Though Neethlingshof has more Pinotage vines than any other red variety, most are used to produce juice for white wines.

to wait virtually till bottling time, sometimes even later, to know whether they've struck a good vintage. Give me our conditions and skill, and their equipment—which we have—and we'll make great wines."

The Neethlingshof Estate makes pure varietal wines from Chenin blanc, Weisser Riesling, Gewürztraminer, Kerner and Bukettraube. The grapes are picked by Gys van der Westhuizen's team at full-ripe stage (23 ° Balling). Schalk crushes them and aims for as much contact between grape skin and juice as possible before the skins are pressed, to get the most richly-flavoured juice. The liquid is chilled to settle out any remaining solid material and then inoculated with yeast and fermented cold (12-14 °C) in closed steel tanks until all the sugar has been fermented out. The wines are racked (the clear wine is pumped away from the inactive yeast settled to the bottom of the tank) and allowed to mature in steel for a short period before being filtered and bottled.

The red wines are also produced by using relatively low fermentation temperatures (20 °C+) and the skins are left long enough to provide sufficient colour, and then removed and pressed.

Neethlingshof has a large harvest of grapes producing a large volume of wine. Most of this wine is made for a wine merchant's use, and less than five per cent is bottled under the Estate label.

Wine is for sale on the Estate.

Overgaauw

Overgaauw's 70 hectares of vineyards are grown on east and south-east facing slopes in the Polkadraai area of Stellenbosch. The land was chosen by Hendrik Elbertsz in 1682, when Simon van der Stel first invited settlers to take up land in the Stellenbosch area. Elbertsz was one of the Cape's pioneer farmers, and took advantage of each opportunity to open up new areas. In partnership with Steven Jansz Bothma, he was granted one of the first farms on the Liesbeek River, in what is now Cape Town's suburb of Mowbray, in 1657. The same year, he joined Abraham Gabbema on an expedition into the interior, and on the way named the Paarl Mountain and the Berg River. Elbertsz was also one of the first Cape burghers to take up the Colony's offer of grazing land in the Hottentots Holland area and when he took possession of By-denweg, six kilometres from Stellenbosch, this ex-midshipman was undoubtedly one of the most experienced farmers in the infant colony.

In 1907, when Abraham van Velden purchased a section of the old farm from his grandparents, Overgaauw was separated from By-denweg. His father had married a daughter of the Joubert family, but both his parents had died comparatively young, leaving Abraham to grow up in his grandparents' care. Abraham gave his portion of the old farm his grandmother's maiden name, Overgaauw, and thus established a place for this old Dutch name in Cape history. Abraham planted vines and built a small cellar to make his first wine. He harvested for his first vintage in 1909 and sold the three tonnes of wine for £30 a tonne. The second season saw a bumper harvest and Van Velden made 30 tonnes of wine, receiving £3 a tonne for it! This was the story of the wine business then. Bad year, good price. Good year, bad price.

Abraham was primarily a white wine producer and most of his vineyards contained Steen and Green Grape (Sémillon). Red wines were made from Cinsaut. Abraham's son, David, started working with his father in 1939, and took over the Overgaauw management in 1945. He planted small vineyards of Cabernet Sauvignon and Shiraz to improve the standard of the Estate's red wines, and a mixed vineyard of Malvasia Rey, Cornifesto, Tinta Barocca, Tinta Francisca and Souzão, to make port. The red wine experiment was not really successful, as wine merchants were unwilling to pay enough to make the production of quality red wines from low-yielding varieties

The Overgaauw Estate, running along the base of the picture, has slopes facing south-east in the Stellenbosch Kloof.

worthwhile, and within a few years the Shiraz vineyard was removed. Although the port vineyard made sweet wines that sold well for many years, the market for ports did not survive the 1960s. David began to make blended red wine from this vineyard by crushing the grapes together and fermenting the must dry. This blend was one of the first Estate products bottled by the family and was originally sold under the name Dry Red. Later, the name was changed to the more colourful Overtinto.

During the last few decades Overgaauw has made white wines from Steen, Clairette blanche, Colombard, and also Sylvaner — one of the first of the special varieties to make a mark on the Cape wine industry. David planted the vineyard in 1959 and made his first wines from it in 1962. The first Estate-bottled Sylvaner wine was sold in 1971 and later vintages caused a great deal of interest in this fresh, bracing style of wine. However, a number of even more promising newcomers were released in the early '70s and producers planted Gewürztraminer, Weisser Riesling and others in preference to Sylvaner, so Overgaauw's product remains unique.

In 1973 David was joined by his son Braam, who had completed a year's training in wine technology at Geisenheim in Germany. Together the Van Veldens have lifted the farm's capacity to produce high quality wines, particularly reds. They started the programme with intensive vineyard redevelopment involving the reconstitution of the soil and the provision of optimum growing conditions for the vine. Fifty hectares of the farm are used to support a herd of beef cattle, which in turn produces waste by-product that is mixed with the skin, stalk and pip residue from the wine cellar. This is accumulated as compost and used for the preparation of a new vineyard. The soil of a new block is analysed and deficiencies are corrected during a period of deep ploughing. The base soil is broken up, the top soil is aerated and thoroughly mixed with the compost, giving the young vine roots every chance to spread deep and wide in search of nourishment.

Overgaauw has a number of large dams and these are used to collect water from winter rains to supplement the vineyards' moisture in summer. All young vineyards are provided with irrigation to help the young plants' growth during summer difficulties.

One of the Cape's top red wines in recent years has been the pure Cabernet Sauvignon made at Overgaauw.

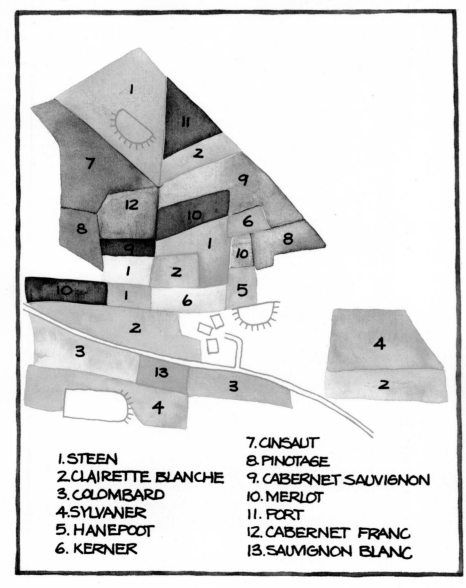

1. STEEN
2. CLAIRETTE BLANCHE
3. COLOMBARD
4. SYLVANER
5. HANEPOOT
6. KERNER
7. CINSAUT
8. PINOTAGE
9. CABERNET SAUVIGNON
10. MERLOT
11. PORT
12. CABERNET FRANC
13. SAUVIGNON BLANC

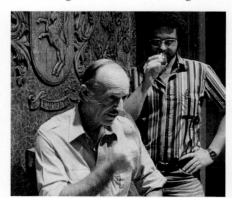

Father and son, David (seated) and Braam van Velden, are producing some of the nation's top reds.

Young vines are planted through plastic sheeting to provide an even better start on Overgaauw's rich soils.

type soil. David and Braam have planted Kerner, a relatively new variety, imported from Germany, on these richer soils. "It's not an easy variety," says David. "It hasn't been terribly impressive, so far, but the vines are getting older and the wine is improving. We'll have to wait and see." Sauvignon blanc vines have also been planted and together with Chenin blanc, Colombar and Sylvaner produce the white wines made in the Estate cellar.

"It's no light matter, deciding which variety to plant," says David. "This farm is very adaptable and I think most varieties will make quality wine here. We have to judge which variety will make reasonable quanti-

During the mid '70s Merlot was planted at Overgaauw, to be followed shortly afterwards by Cabernet franc. The first few years of Merlot production saw a Cabernet Sauvignon, Merlot and Cinsaut blend, which is sold under the name of Tria Corda and is proving to be a success with a market that has traditionally resisted blended wines. Overgaauw has made some exceptionally fine Pinotage wines, and though this cultivar has lost some shine in recent years, it provides a full flavoured red wine from the Overgaauw vineyards that is easy to drink within two or three years of harvest. All of the Overgaauw red vineyards are planted in the rich, dark soil above the house.

The farm has a substantial section of sandy soil and only white cultivars have been planted on this light base. White vineyards outnumber the red two to one, and white varieties have also been planted in the red Hutton-

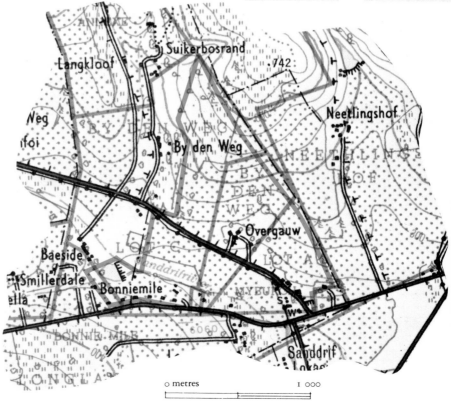

Overgaauw

Overgaauw's young vines are planted with a protective film of plastic over the developing roots. Added moisture provides stronger growth.

The Estate vineyards slope to the south-east, away from the sun. Stellenbosch may be seen in the distance.

Winter vegetation is ploughed into the soil to provide extra humus in the soil structure for the vine roots.

ties of the type of wine that our customers are going to want for the next 25 years, at least. It costs many thousands of rands per hectare to plant a vineyard and each new vineyard will take several years to pay back the investment. If you change your mind or the public don't want to buy a certain type of wine, you'll have to start again." The planting programme at Overgaauw is relatively conservative, with a careful selection of material before propagation and planting. David and Braam look for healthy vines that can produce considerable yields of high quality wine. "Lower yields don't necessarily mean quality," says David. "They could indicate unhealthy vines. We believe that you can have quality and quantity."

David believes the south-east summer breeze that is felt at the top of the farm and the cooler temperatures experienced in the elevated vineyards contribute to the quality of some of the Estate's white wines.

Though Overgaauw is famous for full-bodied Cabernet Sauvignon, the grapes are now being harvested at a slightly earlier stage of ripeness, with pH always below 3,5. The sugar content is lower and the juice is fermented for a shorter period in contact with the skins. The skins and must are separated as soon as the required colour has been obtained. The must is fermented in stainless steel tanks and produces a wine with moderate alcohol and little tannin. This wine is matured in large oak casks for up to two years before bottling. A quantity of small oak casks are being installed to provide an extra dimension to the ageing of pure Cabernet Sauvignon and the Bordeaux-style blend.

The cellar tank press is used to give extra flavour to the white musts. During the harvesting season the press is filled by the morning picking-teams, and the grapes and juice are allowed to soak until 2 p.m. The liquid juice is then run off and the skins are preserved to obtain the rest of the useful juice. The process is repeated in the afternoon. The free run juice and the pressed juices are kept apart, chilled and settled, then inoculated with yeast and fermented at a cold temperature until dry. If required, the wines are blended at this stage.

Owing to a rekindling of interest in fine ports among South African wine lovers, David and Braam have made small quantities of port from their mixed port variety vineyard during some of the recent warmer vintages, when the sugar content of the grapes was sufficiently high. These ports have been matured in oak and are available from the Estate.

Traditionally suppliers of matured wine to a wholesaler, the Overgaauw Estate only bottles about ten per cent of cellar production.

Wine is for sale on the Estate.

Rust-en-Vrede

Situated on the cool slopes of the Helderberg, all of Rust-en-Vrede's vineyards slope north towards the sun.

Grapes picked in lug boxes on this all-red variety farm are being transported to the Estate cellar.

The Estate has south-facing and north-facing slopes, on each side of a deep cleft in the foothills of the Helderberg Mountain, where the Bonte River comes out of the hills on its way to join the Eerste River. Rust-en-Vrede is in a line between Blaauwklippen and Alto and shares with those Estates an ability to make stylish red wines.

Rust-en-Vrede was originally the higher section of Bonte Rivier, and both farms have steeply sloping land that must have been difficult to cultivate in the time before mechanical agriculture. Willem van der Wereld, one of the Cape's earliest farmers, married fifteen-year-old Anna Louw in 1685 and the following year moved on to the land he had been promised by Simon van der Stel. He cleared two long strips of mountainside, one above the other, and was officially granted ownership of the separated pieces under the name Bonte Rivier in 1694. After Willem died in 1696 his widow sold the farm and married Hendrik Elbertsz, who farmed closer to Stellenbosch. At this time, the two pieces of land were still separated and it is unlikely that there were any substantial buildings on either piece. Bonte Rivier passed through the hands of seven different owners over the next fifty years before it became the property of Adrian van Brakel and George Grommet. The Van Brakel family built an imposing house, in the prevailing style, on the lower section of the farm and a large cellar on the upper section, both erected around 1780. In 1801 Bonte Rivier's two pieces of land were bought by Johan Liebetrau, who was to build a major new cellar near the house on the lower portion and a new house beside the old cellar on the other portion. On Liebetrau's death in 1832 the property was divided between his sons and the upper portion became known as Rust-en-Vrede. The remaining section, Bonte Rivier, was sold in 1846 by one of the Liebetrau sons to Jan Hendrik Hofmeyr, the distinguished father of the distinguished politician and statesman of the same name.

In addition to the 1780 cellar and 1825 house, Rust-en-Vrede has a Jonkershuis dating from 1780 and a ruin of another cellar that has been estimated to date from 1730. Vineyards were planted and wine made on the farm from the first settlement. Willem and Anna van der Wereld were stated to have 7 500 vines in the 1692 Cape census, and wine was made on the property for the next two hundred years. But the cellar fell into disrepair through neglect during the present century, and only when Rust-en-Vrede was bought by Jannie Engelbrecht in 1978 did the spirit of restoration animate the farm. During 1978 Jannie replanned his new 50 hectare farm and assessed each of the old vineyards. The house needed attention and the cellar was being used as a stable. The grape harvest from the vineyards was sold to a wine merchant. Jannie decided to follow the lead of neighbouring Alto and make Rust-en-Vrede an exclusively red wine Estate.

Jannie grew up on his father's grape-growing property between Vredendal and Klawer on the Cape west coast, where Hanepoot vines were cultivated to make raisins. After Stellenbosch University, a degree in economics and a career in rugby, Jannie bought a farm at Koekenaap, not far from Vredendal, where he planted vines and grew cash crops of vegetables. In 1970 he sold the Koekenaap farm and concentrated on his share of the Vredendal land. With a background in viticulture and Stellenbosch rugby, Jannie yearned to return to the heart of Western Province's wine territory and began to look for a small Stellenbosch Estate where he could make quality wine. Rust-en-Vrede's rich Hutton and Clovelly soils gave him the opportunity to indulge his ambitions.

After the first harvest, Jannie removed some of the less productive of the old vineyards and planted red wine varieties in their place. During 1979 and 1980 he made white wine from the remaining Chenin blanc vineyard and red wine from the Cabernet Sauvignon, Shiraz and Tinta Barocca vineyards that were already in production when he bought the farm. In 1980 the remaining Chenin blanc vineyard was removed to make way for more red varieties.

Rust-en-Vrede is on the western slope of the Helderberg, and so the sun rises late, but as almost all the vineyards, sheltered from summer winds, face north, they get warmth throughout the rest of the day. With

Rust-en-Vrede's vineyards face a deep cleft cut by the Bonte River.

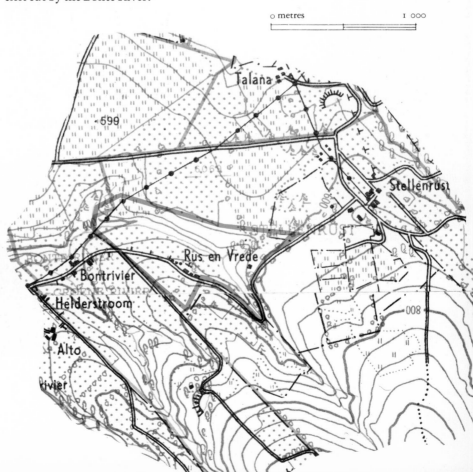

The Estate is on the lower slopes of the Helderberg, just north of Alto.

174 Stellenbosch

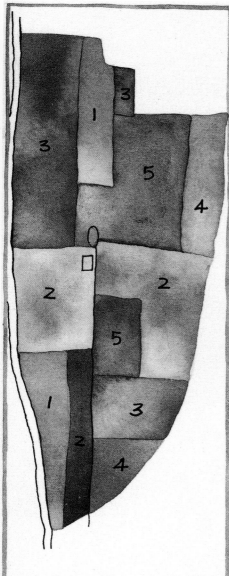

1. PINOT NOIR
2. CABERNET SAUVIGNON
3. SHIRAZ
4. TINTA BAROCCA
5. MERLOT

clay-rich soils and enough natural moisture, early- and late-ripening red wine varieties are able to ripen without stress and produce grape juice with enough sugar and acidity to make well-balanced wine with rich colour.

After an analysis of the climatic influences on neighbouring Alto's red wines and the results of his first crop of grapes, followed by his own

maiden vintage of wines in 1979, Jannie's planting programme has centred around further vineyards of Cabernet Sauvignon and Shiraz, with additional plantings of Pinot noir and Merlot. Because of the size of the farm and his own consuming interest in red wine, he is aiming for the top of the market by producing wines of the highest quality.

Jannie has opted to forgo the potentially greater quantity of wine that he could get by irrigating his vineyards with water from the Bonte River, and has a strict no-irrigation policy. Vines are pruned short to reduce the crop, and the excess bunches are cut off in early summer. He aims to grow healthy vines, with a restricted crop of fully-ripe grapes that can be harvested at the chosen degree of ripeness to make medium-bodied deep-coloured wines with the capacity to last for many years. The grapes are harvested mid-ripe (21-22° Balling), to get high acidity with moderate alcohol, crushed and then fermented at cool temperatures in stainless steel closed tanks. The skins are kept in the fermenting must until about half the sugar has been fermented out and sufficient colour has been extracted, and are then removed and lightly pressed to get the last of the quality liquid. After fermentation is complete, the new wine is racked to remove most of the lees in the base of the tank and then left for malolactic

All of Rust-en-Vrede's vineyards slope to the north and west and are planted with red wine varieties.

Rust-en-Vrede

fermentation to begin.

The Rust-en-Vrede style of red wine requires thorough maturation in wood and Jannie has installed enough small and large oak vats to mature his whole crop. As new vineyards bear their full crop, increasing the quantity of wine made in the cellar, the quantity of oak-maturation vessels will be increased to cope. Jannie believes in causing red wine to achieve its greatest flavour potential from the grape, and then giving each wine sufficient wood and bottle maturation to be able to assess accurately the merit of the wine and to get the maximum enjoyment from each bottle. To be able to give as much of his crop as possible the required optimum maturation, he sells a small proportion of each vintage relatively young and the balance stays in storage.

The red wines are kept in oak long enough for them to acquire an extra degree of character. The length of this period depends on the vintage, the cultivar and the size and age of the maturation cask. All Rust-en-Vrede's top red wines have part of their maturation in new oak and also in small oak casks. Cabernet Sauvignon from a warm ripening season will spend six to eight months in small vats and up to two years in 4 000-litre vats. Wines from cooler years will have shorter maturation periods. During their time in wood the wines are tasted from time to time to assess the degree of maturation and the extra character that has been obtained from the contact with oak.

After bottling the wines are stored for a further period of bottle maturation. Jannie Engelbrecht's policy of retaining as much of his wine for as long as necessary to give it optimum maturation before sale will undoubtedly result in the availablity of wines of different vintages at Rust-en-Vrede.

Wine is for sale on the Estate.

Relatively small for a Stellenbosch Estate, Rust-en-Vrede allows the Engelbrecht team to give particular attention to the vines.

With the bulk of the Helderberg to the east, Rust-en-Vrede's vineyards have a late sunrise each day.

Looking across Rust-en-Vrede's vineyards during the harvest, toward the slopes of the Stellenboschberg.

Schoongezicht/Rustenberg

When Roelf Pasman selected the farm Rustenberg (from which Schoongezicht was later separated and to which it was eventually reunited) in the foothills of the Simonsberg in 1682, he laid the foundations of a dynasty that made many of the great wines of the early Cape. Pasman married Sophia van der Merwe in 1684 and a daughter, Sibella, was born of the union. After Pasman died in 1695, Sophia inherited the farm and married Pieter Robbertsz, who was later to become acting-Landdrost of Stellenbosch, the most important official in the region. Robbertsz was officially granted Rustenberg in 1699 by W. A. van der Stel and was one of the Governor's chief supporters during the revolutionary period of 1705-6.

In 1714, Sophia bought Nooitgedacht as an inheritance for Sibella, and she lived there with her husband, Johannes Loubser, and their family. One of her children was Pieter, who inherited Rustenberg in 1742 on the death of Robbertsz. After Johannes Loubser's death, Sibella married Jacob Cloete. Nooitgedacht was to remain an important wine-producing property during the Cloete period of ownership, which lasted for 77 years. Sibella Pasman's grandson, Hendrik Cloete, was assisted by his father to buy Groot Constantia in 1778, and became the Cape's most famous wine producer.

Rustenberg became the property of Jacob Eksteen, one of Sibella's descendants and a cousin of the Loubsers, in 1786. Eksteen's daughter and her husband Arend Brink also lived on the farm, and in 1800 Brink built a cellar next to his house on part of the property. These two buildings are still in use as home and cellar today. In 1810 Eksteen formally transferred this tract of land, which became known as Schoongezicht, to Brink, shortly before both father and son-in-law sold their properties. Schoongezicht was bought by another Hendrik Cloete, a grandson of the aforementioned Hendrik Cloete, and

Cool grapes are valued in Cape cellars and harvesting commences before sunrise on Schoongezicht, to obtain the day's crop before the peak temperature.

Schoongezicht was separated from Rustenberg in 1810 and the two farms were re-united in 1945.

the farm was to remain in Cloete hands until 1850.

In August 1892 John X. Merriman, Treasurer of the Cape Colony, was appointed Minister of Agriculture, and he at once decided that, in order to know what he was talking about, he should himself become a farmer. At that time the biggest problem facing the wine farmers was the scourge of phylloxera, which had already wiped out the vineyards of Europe and was doing the same in the Cape. In order to associate himself with those most seriously affected, he bought the farm Schoongezicht, whose vineyards were completely derelict. He did not have much capital, but at that time farms were going for a song — in 1886, six years before he bought Schoongezicht, he wrote in a letter to J. B. Currey: "By jove, I wish I had a few thousands lying idle or even hundreds. Land at Stellenbosch and Paarl is just being chucked away. Haupt's place 'Rustenberg', the finest place in the district, two hundred thousand vines — first mortgage £4 000 — sold on Wednesday for £2 400!"

Rustenberg was the farm from which Schoongezicht had been excised in 1810, and two years after Merriman bought Schoongezicht, his sister Charlotte, wife of Sir Jacobus Barry, bought Rustenberg. Even before the phylloxera, Merriman had felt that South Africa was making inferior wine. "Look at our vineyards," he wrote to J. B. Currey in 1887, "a mine of wealth at our very doors. Yet while Australia is actually making a claret which fetches eighty shillings a dozen, we are selling our filth for £4 for one hundred and twenty seven gallons, and struggling for a protective excise to enable us to do that."

He was joined on Schoongezicht by an 18-year-old immigrant, Alfred Nicholson, who later married his niece and became his partner in the farming enterprise. Merriman and Nicholson set to work at Schoongezicht and rooted out all the vineyards. Where vines had been, peaches, plums and pears were planted. Land for vines, hand-delved, was prepared on virgin, bush-covered hillsides, and good cultivars, grafted on phylloxera-resistant rootstock, were obtained.

During his early years at the Cape, Merriman had sold wines imported from the great Estates of France and Germany, and now with Alfred Nicholson's help he seized the opportunity to try to make Cape wines of a similar quality. When Schoongezicht's orchards, vineyards and cellar were in full production, Merriman turned to Europe for markets and was one of the first to pack and ship fruit to Britain. He persuaded the London firm of Burgoyne's to take an interest in Cape wine (they were already importing Australian wine), and soon Schoongezicht wines were being sold throughout Europe. Burgoyne's were later to add Alto and Twee Jongegezellen wines to their price lists.

Most of Alfred Nicholson's life on Schoongezicht was spent in healthy competition with the Barrys on the adjoining farm. When the elder Barrys died, their sons sold Rustenberg to Baron De Villiers, who had been President of the South African National Convention, the body that had thought out the framework of what became the Union of South Africa.

When the Second World War began, and trade decreased, Reg Nichol-

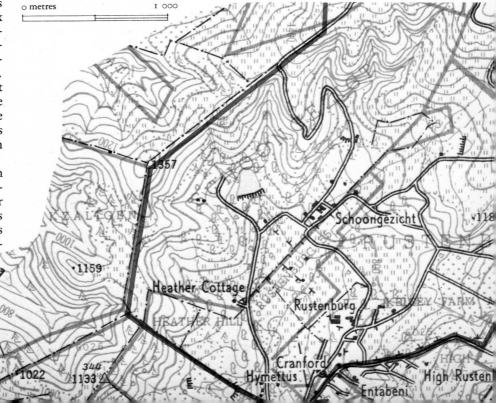

A. NICHOLSON, SCHOONGEZICHT, STELLENBOSCH.

Jersey cattle today graze where the farm's original vineyards were planted. (left) Alfred Nicholson the first Estate vintner, shown promoting the farm's wine.

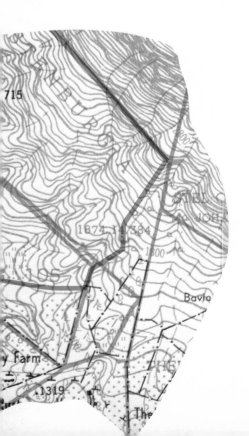

son, Alfred's son, could see no other answer to his farm's troubles than to sell. The renovation of the cellars and the construction of dams in spruits on Simonsberg's higher slopes for use in times of drought needed cash, which the Nicholsons could not raise. In 1940 the Barlow family had bought Rustenberg, and Reg Nicholson had become friendly with Peter Barlow, a member of the family. A few years later Peter Barlow was fighting on the North African front, when he was told that Schoongezicht was for sale. Somehow he managed to get word back to his family that they must dissuade Reg Nicholson from selling to anyone other than the Barlows, who would reunite the old farms.

In 1945 Peter Barlow bought Schoongezicht and was subsequently able to carry out improvements to both Schoongezicht and Rustenberg. This at last allowed Reg Nicholson, a graduate in agriculture from Stellenbosch University, to tap the natural resources of the land and the climate to make wines that could earn a place in South African history.

The Schoongezicht cellar was chosen for renovation, future production and bottling; the old Rustenberg cellar is unused. Both farms carry expensive and famous dairy herds on the lower, flat lands, where kikuyu grass, rye and clover grow in profusion. Standing in these fields, looking towards the mountain, one sees the bright green of grass, the darker leaves of the oak trees and hedges, and the magnificence of Simonsberg. Here, on the flat, where the first Rustenberg vines were grown, there is now only irrigated pasture. The vineyards have moved to the lower slopes of Simonsberg, and new planting simply moves further up the mountain.

The names at times confuse, for white wines have always been marketed under the Schoongezicht label, and red wines under that of Rustenberg; but both are from the same cellar.

Wine has been grown, made and bottled on the combined farms from the early 1900s. Wine was made on Constantia, Alphen and several other farms before this date, but Schoongezicht seems to be the only farm to have continued with the practice of bottling wines from the early years of this century until the present day.

The date 1800 appears on the gable of the farm's cellar, for the original building is still being used. Many additions have been made to the cellar since then, most of them without record. When recent additions were made, necessitating the tearing down of adjoining walls, some of them were found to contain just rock and river sand, others bricks and a rough sand mortar. The eastern end of the old cellar is approximately seven metres longer than the end facing the house, and the front gable and the back gable are some two metres out of line. Very little of what we can see now as the Schoongezicht cellar was built at the same time. The present owners have plans to extend the cellar further, and these additions will be made to harmonise with the existing buildings.

In recent years, the area under vineyards has increased and the historic cellar has been unable to provide sufficient space for maturation. An additional cellar will house oak maturation casks and enable Schoongezicht

Schoongezicht/Rustenberg 179

Schoongezicht's cluster of buildings comprises the Estate home, stables, dairy, winery and maturation cellar.

to turn more of the produce of its own vineyards into quality wines, bearing the familiar labels of Rustenberg and Schoongezicht.

High on the slopes of the Simonsberg, Schoongezicht has vineyards facing east, south and south-west, growing in decomposed soils of the Hutton type. These mountain soils contain a high proportion of clay and retain moisture throughout the summer. Schoongezicht and its sister farm are in the shade of the Simonsberg and receive eighteen hours less direct sunlight per month than farms in the centre of the valley. Attracted by the bulk of the mountain, moist air condenses and provides Schoongezicht with 900 mm average annual rainfall, more than the average for the area. The extra rain, less direct sun and the moisture retention of the soil combine to assist the growth and allow the optimal ripening of the late-ripening cultivars.

Though some early and mid-season white cultivars have been planted to make use of the favourable ripening conditions, the majority of the vineyards contain late-ripening red varieties. Traditionally, the great red wines of Rustenberg have been pure Cabernet Sauvignon or Cabernet-based red wines, but Etienne le Riche, the Estate vintner, has planted, in addition to more of his premier variety, blocks of Cabernet franc and Merlot to expand the selection of reds for blending purposes.

Although the Rustenberg wines have always been predominantly Cabernet Sauvignon, in 1968 a new fixed pattern of blending was initiated by crushing Cabernet Sauvignon and Cinsaut grapes together in the proportion of two to one. This wine became the Estate's basic product and has remained so. Until 1973 this blend was known as Rustenberg Dry Red, but from the 1974 vintage onwards the last two words were dropped from the label. The Cabernet franc and Merlot grapes will be used to make wines to blend with Cabernet Sauvignon and provide the Estate with a second blended red.

Schoongezicht's large Cabernet Sauvignon vineyards are also used to produce two varietal wines. Some of the Cabernet Sauvignon must is drawn off the fermenting mash as soon as it has obtained a deep pink colour and is fermented dry in closed steel tanks to produce a full-bodied, dark-coloured rosé. Other Cabernet Sauvignon grapes are picked full-ripe (averaging 22° Balling) to make a full-bodied wine. After crushing, the skins are left in the fermenting must, with the temperature held below 27 °C, while the yeast from the bloom on the grapes ferments the sugar into alcohol. Once sufficient colour has been obtained (a longer period in a cooler year) the skins are removed and the wine is allowed to ferment dry.

Pinot noir is the only early-ripening red cultivar in the Schoongezicht vineyards and Etienne le Riche has had more experience with this difficult noble variety than most other Cape cellar masters. The vines have a tendency to overbear and Etienne has

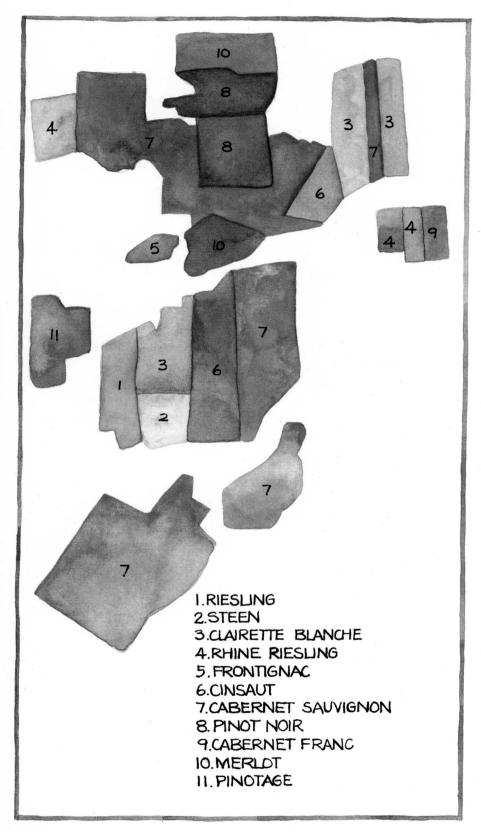

1. RIESLING
2. STEEN
3. CLAIRETTE BLANCHE
4. RHINE RIESLING
5. FRONTIGNAC
6. CINSAUT
7. CABERNET SAUVIGNON
8. PINOT NOIR
9. CABERNET FRANC
10. MERLOT
11. PINOTAGE

pruned them as short as possible. He believes that Pinot noir requires longer fermentation (at lower temperatures) with the skins in the mash than Cabernet Sauvignon. It also requires wood maturation, and when successful provides a ruby red wine, and an intensity and complexity of fruity flavour that is unrelated to colour depth.

To maximise the quality of the wines made on the Estate, Etienne has planted vineyards with a high vine population per hectare, expecting a low quantity from each vine. He tries to keep the wines made from older vines separate from those made from younger vines, believing that more mature and deeply established vines produce superior grape juice.

Thirty per cent of the Estate vineyards contain the white wine varieties Cape Riesling, Clairette blanche and Weisser Riesling, all mid-season to late-ripening varieties, and Chardonnay, the sole early-ripener. The grapes are picked in lug boxes (which are used for the whole crop), crushed and left for half an hour, with juice and broken skins in contact. The juice is separated by squeezing the husks in a press, then inoculated with a yeast strain and fermented at cool (15 °C) temperatures for between three weeks to a month until all the sugar has been converted to alcohol and carbon dioxide. After cleaning, some of the Weisser Riesling is matured in small oak casks to obtain an extra degree of character before bottling. The dry wines made from the other white varieties are stored in steel tanks until ready for the bottling line. All Schoongezicht's wines are given a period of bottle maturation before release to the market.

Finally, the Estate has a small vineyard of Muscat de Frontignan, which is currently used to make Schoongezicht's fortified dessert wine, a product first popularised by Alfred Nicholson.

The Estate's white wines are sold under the Schoongezicht label, while the red wines are sold under the Rustenberg label.

Wine is for sale on the Estate.

Simonsig

After the white grapes are crushed, the skins are left in the juice to gain extra flavour. The juice is then drained off.

Looking to the south across the Simonsig cellar to the north-facing Simonsig vineyards four kilometres away.

The wine in these bottles is fermenting, producing the characteristic bubbles of sparkling wine. The yeast will be shaken down and removed.

The Simonsig Estate consists of two separate farms, Simonsig and De Hoop. Simonsig has 90 hectares of vineyard on north-facing slopes just outside Stellenbosch while on De Hoop 86 hectares are planted with vineyards on south-east-facing slopes. These farms constitute one Estate, as they were joined before the Wine of Origin legislation was passed in 1973.

The southern farm has richer soils and a steeper slope with greater variation in temperature, but the differences between the two properties are not great. The moderate Cape climate and the use of irrigation and other progressive farming techniques allow these two farms to provide the Simonsig cellar with a sufficient diversity of grapes to make one of the largest ranges of quality Estate wines in the Cape.

The two Simonsig farms were originally part of different old Cape properties. The southern farm was once part of Nooitgedacht, granted to Matthys Greef, a blacksmith, by Simon van der Stel in 1682. Greef arrived in the Cape in 1680, bought blacksmith tools from a smith in Cape Town and was promised the attractive farm on the Kromme Rhee River, where he was to establish the base of a successful farming business. By 1692, he had 10 000 vines and grew grain at Nooitgedacht, while he leased land for sheep grazing in Drakenstein, Tulbagh and Malmesbury.

Greef understood the art of making medicines and plasters from herbs, with which he treated animals, slaves and anyone who needed help. He was very well-to-do and in 1709 bought a big house in Cape Town. He died in 1712 and the farm soon became the property of the Pasman family who owned Rustenberg. Sibella Pasman inherited Nooitgedacht and in turn bequeathed the farm to her son Hendrik Cloete, whose son was to move to Groot Constantia and make the wines of Simon van der Stel's Estate world-famous. The Cloetes owned Nooitgedacht for more than a century.

Simonsig's other set of vineyards was part of the historic Koelenhof

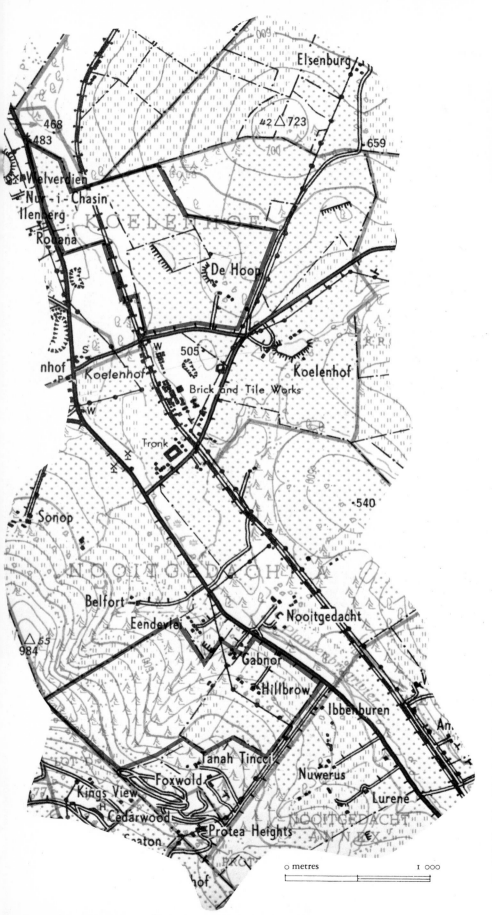

Frans Malan, dedicated activist in the production of quality wines, runs Simonsig with the help of his three sons.

farm, promised to Simon de Groot in 1682. It is situated on almost level ground in the centre of the valley between the Simonsberg and the Bottelary Hills.

The adjoining properties were first brought together in a grand scheme to develop the quality of the Cape wine industry and build a secure export market in Europe. The first plan, in 1888, proposed to buy top wines from Cape farmers for export sale. When this scheme foundered, Dr. Theophilus Hahn proposed the formation of a syndicate to purchase Nooitgedacht, Koelenhof and four or five surrounding farms believed to be the cream of Cape wine producers, for the establishment of top-quality vineyards and a modern cellar that would produce European-quality wines and would in turn make a lot of money from exports. In 1889, when the first sign of phylloxera was noticed in Cape vineyards, the project was abandoned. Most Cape vineyards disappeared during the phylloxera attack. Nooitgedacht and Koelenhof were united under the ownership of Cecil John Rhodes and orchards replaced the stricken vineyards.

In 1953 Frans Malan bought the farm De Hoop, a portion of Koelenhof, from his father-in-law and began to make wine in the existing cellar. Ten years later, Frans purchased a portion of Nooitgedacht, named it Simonsig, and brought the two into one operation. The southern farm was almost undeveloped at the time of the purchase and cultivars had to be chosen for planting.

During the early '60s, the Cape wine industry was under the influence of two new developments, rapid growth in the sales of low-priced white wines and a dramatic improvement in the produce of cellars using

Simonsig

cold fermentation techniques. Frans, a young and dynamic farmer, made wine for sale to merchants, and like his colleagues was encouraged to produce large volumes of white wines of basic quality that were destined to be sold in the popular-price market. For greater success in this market, Frans planted mostly bulk-producing white wine cultivars and built a new cellar with refrigeration facilities in 1964. The overall quality of the wines made in the Simonsig cellar improved with the new technology, but Frans began to realise that he would have more influence over his own business and a greater chance of financial success if some of the cellar's best wines were bottled and marketed under the farm's own label.

In 1968 he bottled 600 cases each of Chenin Blanc, Cape Riesling and Clairette Blanche wines and sold these through direct mail. The success of his initial marketing operation changed the course of the Simonsig Estate. He began to explore the possibility of bottling and marketing more of his crop, and together with Niel Joubert of Spier and Spatz Sperling of Delheim developed the concept of the Stellenbosch Wine Route.

It is not an easy matter to turn a wine farm producing bulk wines into a fine wine Estate. The major difficulty lies with recently established vineyards, stocked with standard varieties, producing an important share of the cellar income, since a wine Estate requires special varieties, producing premium-quality wines.

At that time, Frans's two farms were planned to produce large volumes of wine with a consistent standard of quality at the lowest possible price. These wines, like those of Frans's neighbours, were bought by wholesalers to be resold under nationally advertised brands. The vineyards supplying these wines and the cellar philosophies that went with them did not easily fit with the new style of Estate wine marketing.

The costs of establishing and maintaining a cold fermentation cellar are high, while the yields of the vineyards in the Stellenbosch area are compara-

A Stellenbosch morning mist lies across the Simonsig Estate as we look over the vineyards toward Simonsberg.

Simonsig's soils, of poor to moderate fertility, provide vineyards with sufficient growth to ripen a reduced crop.

Each summer most of Simonsig's vineyards receive supplementary irrigation, supplied from water stored in large dams.

tively low. Substantial profits need to be calculated into the production of small-scale wineries to make them worthwhile. The cultivars grown by Simonsig, like those on most other farms, belonged to the wrong quality bracket and Frans started replanning his vineyards.

He was the first Cape farmer to show an interest in newly released Weisser Riesling, Gewürztraminer, Bukettraube and Kerner grafting material, obtained and grafted out to rootstock in 1971 and planted in Simonsig vineyards in 1972.

Frans had already embarked on a scheme to provide his Estate with a flagship, a prestige product made by an age-long technique that Cape wine producers had abandoned in quest of greater turnover many years before. Champagne bottles were imported from Europe, riddling racks were bought from France and Simonsig went into the bottle-fermented sparkling wine business in 1971.

However, the improvement of the content of Simonsig's vineyards was

The Simonsig Estate cellar produces top wines, as well as a Méthode Champenoise sparkling wine.

the major task ahead and the first Cabernet Sauvignon vineyards were planted in 1973, followed by Shiraz and Pinot noir. Though the quality of the early Chardonnay vines was suspect, Frans planted a small vineyard when the vines became available to see what this classic variety could produce.

Most of the southern farm and a part of the northern farm had been planted shortly before the decision to swing from quantity to quality vineyards. Owing to the high cost of replanting and to maintain supply for the ever-increasing demand, the scheme had to be implemented in stages.

By 1973 Frans was marketing ten per cent of his crop as Simonsig Estate wine and the rest was being sold in equal shares to two wholesalers. In that year, both wholesalers declined to buy his wine and a whole new strategy had to be devised. He obtained a wine farmer's licence, enabling him to sell the products of the Estate at other premises, and in addition to selling Estate wine from the farm he became a one-brand retailer of low-priced wine in Stellenbosch and, later, nearby Kraaifontein.

This stop-gap tactic ran alongside a programme to improve the standard of the vineyards and gradually bring more of the Estate's produce into bottles bearing the Estate label. Whereas climate and growing conditions are normally the factors that influence the types of wines that an Estate produces, the need to sell large quantities of wine to interested and loyal consumers to maintain the financial health of 170 hectares of vineyard has caused Simonsig to exercise ingenuity in the development of new styles of wine.

Through the '70s, the Estate had more Chenin blanc vines than any other variety and made dry and semi-sweet wines from this versatile but standard-quality cultivar.

In 1978 Frans experimented with the production of a wood-matured white wine using Chenin blanc and cautiously introduced a trial batch of this uniquely-flavoured product. The sales encouraged greater production of the wine, which was at first known as Fumé Blanc, and has since been renamed Vin Fumé. Today the wine is a blend of Chenin Blanc and Sauvignon Blanc and has become the Estate's top-selling product. The grapes are picked ripe and the juice is fermented dry. The wine is matured in small oak casks until it has obtained an extra dimension of flavour, reminiscent of vanilla, and then bottled and sold within a year of the harvest. This wine, interesting while young, improves greatly with a year or two of age in the bottle.

A parallel programme launched in 1979 produced small quantities of Noble Late Harvest, which simultaneously became a classified, descriptive South African term for a high-priced, after-dinner wine. Small quantities of this wine, generally made from richly-flavoured Weisser Riesling and Bukettraube grapes, have followed every year since then.

One of the first Cape vineyards to plant the American selection of Gewürztraminer, Simonsig has had notable success with this variety. The first Simonsig Gewürztraminers were made off-dry but recent vintages have been produced in the Special Late Harvest category and have a pronounced fullness and much more sweetness. Simonsig semi-sweet wines are made by arresting the fermentation of chosen tanks of wine at the required degree of sweetness. The wine is chilled and passed through a centrifuge. It is then fined and filtered before bottling. These wines are normally cleaned and bottled during winter, taking advantage of the lower temperatures.

The Simonsig Estate has 18 varieties

This red soil, known as Hutton, contains a quality of clay and retains moisture, helping late varieties to ripen fully.

The Simonsig bottling line is closed during the harvesting season because of the air-borne yeast in the cellar.

producing grapes for the cellar, and more than 50 per cent of the wines made in the cellar are sold as Estate wine. The proportion of red to white in the vineyards is about 30 to 70, and most of the medium- to high-quality white varieties are used to add varietal differences to the spectrum of dry and semi-sweet white wines.

Frans has planted Bukettraube, a variety with a subtle muscat character, in some of the Estate's lightest soils, which give the resulting wine a greater freshness of flavour and delicacy of aroma. In contrast, the Chenin blanc vineyards are planted in the heavier soils, where the vines are able to extract the fruity flavour that the wines require.

The Estate programme is heavily weighted toward the production of white wines with distinctive character. "I believe that the varieties make the difference," says Frans. "When we have the right mix of the ideal material, we'll make white wines with character that will surprise the consumer."

The Simonsig red vineyards contain Cabernet Sauvignon, Pinotage, Shiraz, Pinot noir and Merlot, which are used to produce a range of pure cultivar and blended wines. The Simonsig red wines can generally be described as medium-bodied and are deliberately made in the lighter, fruitier style. Most of the red vineyards are

The peaks of Simonsberg stand behind the gently sloping vineyards of De Hoop.

planted on light soils and their grapes are picked at mid-ripe stage. Skins are removed from the fermenting must as soon as sufficient colour has been obtained and the liquid is allowed to ferment dry. When the new wines are clean, they are placed in small oak barrels for up to six months, then transferred to large oak and generally receive a total of eighteen months' maturation before release.

The three sons of Frans and Liza Malan have entered the business and have provided a modern technological base for a new period of development. The eldest son, Pieter, has taken control of the administrative function of running a production and marketing business. The second son, François, is the technical director of the Estate. Johann, the youngest son, is today in charge of the cellar.

Today the Simonsig team markets and nationally distributes almost all the Estate's large volume of wine. A most encouraging sign for the winelover has been the steady improvement in quality alongside the constantly increasing production. We can continue to expect some very interesting wines — particularly whites — under the Simonsig label.

Wine is for sale on the Estate.

Spier

When Simon van der Stel became Commander of the Cape in 1679, he changed the system of land grants and made farms available to almost any man willing and able to endure the hardships of living in the interior and turning bush and shrubland into a viable farm. In the years before the arrival of the Huguenot refugees from France, the majority of land grants were made to German artisans and soldiers who were freed from their indentures to establish small farms.

In terms of this new system, Arnout Jansz, a soldier who had arrived from Germany in 1683, was granted a farm in the Stellenbosch district amongst three other farmers who had settled along the Eerste River the previous year. He called his farm Spier, built an *opstal* on the other side of the river, and started to support himself and his wife, Hendrina van Ysel. These hardy pioneers were lent implements and tools by the administration, and with axe, chisel and plough set about the slow and unrewarding task of clearing land and planting crops. When Jansz had his promissory grant made official in 1692, the land consisted of six small separated pieces, totalling 34 hectares. Presumably each piece was land he had been able to clear and farm. In 1692 Arnout (who was known as 'Tamboer') and 'Henderientje' had three sons and a daughter and had 2 000 vines in the ground.

Tamboer Jansz died in 1706, and his wife sold Spier to Hans Hatting, another German immigrant, in 1712. Hatting had been granted La Motte (today also a wine Estate) in the Groot Drakenstein Valley by Simon van der Stel, but he moved to Spier to live when he bought the farm. Hatting's son-in-law, Johannes Groenewald, inherited the farm, and after his death the farm passed to his wife Clara, who subsequently married the owner of Libertas, Johan Bernard Hoffman, in 1754. In 1765 Hoffman sold Spier to Albertus Myburgh, who built the wine cellar behind the present Spier manor house and the stable, where Spier wines are sold today. For the next 120 years Spier was owned

The original stable building on the Spier Estate has been converted into a tasting and sales facility.

Some of Spier's best soils are on the Goedgeloof section of the Estate, near the cellar. The land slopes to the north and east and provides the vineyards with ideal ripening conditions.

by the Van der Byl family, who became wealthy and built most of the historic buildings that may be seen there today. In 1824 it was recorded that Andries van der Byl had 80 000 vines.

Then followed a century of decline, and when Niel Joubert — whose family have lived in the immediate vicinity of Spier since 1784 — bought the farm in 1965 it was in a neglected condition. He restored all the buildings and consolidated Spier with his other farms to form one wine Estate supplying grapes to the Spier cellar.

The Spier Estate is a 350 hectare unit comprising five adjoining farms. Its vineyards stretch over the north-facing slopes on Polkadraai, over the crest of some low hills and down the east-facing slopes into the Eerste River Valley, where the gabled Spier homestead borders on the riverside vineyards. Spier has become one of South Africa's largest producers of Estate wines, with a wide range of products, ranging from dry whites and reds to ultra-sweet Noble Late Harvest and fortified wines.

The situations and micro-climates of the vineyards and the soils in which they stand vary greatly on this narrow, long farm. Chris Joubert, Niel's son and the Estate vintner, has vineyards of the same varieties planted in different sections of the farm, where they are able to ripen at varying intervals and give different styles of wine. "When we make up a blend, we like to use wines made from grapes grown

Looking across a vineyard planted on a southerly slope toward the lower section of the Estate, beside the Eerste River, where the wine sales and restaurant are situated.

on the same slopes," says Chris. "It gives a particular style to the wine and allows us to maintain the style from vintage to vintage."

The Spier vineyards contain more white varieties than red. Chris is aiming to have a ratio of 70 to 30, matching the South African consumer's current preference. The white varieties include Chenin blanc, Cape Riesling, Colombar, Sauvignon blanc and Bukettraube. The red vineyards contain Cabernet Sauvignon, Shiraz and Pinotage.

Spier has done more to demonstrate the flexibility and under-rated quality of the Colombar grape than any other winery in the country. Chris Joubert allows his Colombar grapes to ripen to 21,5° Balling, to make full-bodied, dry table wines that mature with a broad, complex flavour after several years in the bottle. Alternatively, he allows vineyards with *Botrytis cinerea* infection to ripen their wrinkled grapes to rich sugar concentrations and makes Special Late Harvest and Noble Late Harvest wines from their succulent juice. "If you want a wine to last, it should come from the best areas on the farm," he says. "I planted the Colombar in some of my best soils because I want depth and

The long, narrow Spier Estate runs down the south-facing slopes of the Polkadraai hills, providing a multitude of micro-climates for the growing of many cultivars.

Unlike European wine producing areas, where Estate owners and workers live together in villages, labour on South African Estates must be provided with housing on the farm.

A large proportion of the Estate's vineyards are irrigated in summer and water has to be collected from the winter rains and stored for use during the dry season.

vitality in the flavour of the wine."

Chris has had similar success with Pinotage and uses the variety to make dry red wine, a dry white (Blanc de Noir) and port. The dry red Pinotages have been very successful full-bodied wines, deep in colour, without the distracting aromas and earthy taste often associated with this grape. Most of the Estate's Superior certified wines have been late harvests made from Chenin blanc or Colombar grapes. These, however, have been limited quantity products, and it is the dry white wines that represent the greatest volume of Estate sales.

The moderate climate that Spier, being on top of a substantial hill and relatively close to False Bay, experiences, and the wide range of micro-climates and soils in the vineyards, are probably factors in the development of this Estate's unique style. "Our wines are relatively slow to come forth," says Chris. "They don't have too much nose when they're young. But that doesn't worry us. We can recognise potential for maturation." He sets out to make full-bodied wines. "A thin wine has never become a great wine. I like to have a wine that you can taste, that has flavour and interest in it. We pick our grapes when we feel they're fully ripe, and we try to make sure that we have the vines growing in the type of soil that will allow them to hold plenty of acidity as they ripen properly."

Spier has lately imported 16 000-litre vats from France for the maturation of its red wines. These vats raise questions about maturation technique. The traditional cask used for the maturation of red wines in Cape cellars holds about 4 000 litres. In recent years there has been a dramatic change to follow the French and American lead and many cellars have imported small oak casks, mostly of 225 litres, to give a more distinctive maturation character to both red and white wines.

"I've had experience with small wood," says Chris, "and I'm not happy with it. We don't have any trouble getting tannin in our red wines. What we need is a controlled period of maturation in our warmer

Niel Joubert was one of the founders of the Stellenbosch Wine Route and is a prominent member of the Cape wine industry.

Spier's group of historical buildings includes two homes, two restaurants and the Estate tasting-room, shown here.

The western end of the Spier Estate is in the shallow valley of Stellenbosch Kloof.

temperatures, and we'll get that more reliably in large wood. In the south of France, they have similar climatic conditions to ours, and they use large wood. In the cooler areas, in the north of France, they use small wood to add extra character to the wine. I don't think we need extra elements in our red wines. We mature our reds in wood to make them soft, more easy to drink."

For greater individuality and to assist the development of Spier's style of wine, Chris prefers to ferment all his wines for extended periods at low temperatures. The white wines are generally fermented in the area of 12 °C to retain all the delicate flavours and aromas of the fully ripened grape. The red musts are fermented in closed tanks at around 20 °C. The skins are removed at approximately 10 ° Balling and the must is fermented out slowly at a cool temperature. "I believe that this is one of the reasons why our Pinotages have matured so well in the bottle," says Chris. "The wines can still be fresh after ten years." When fermentation is complete, the wines are put into large wood for a year to mature before bottling. A further period of bottle maturation follows to enable the late-developing Spier style to become evident.

Wine is for sale on the Estate.

Uiterwyk

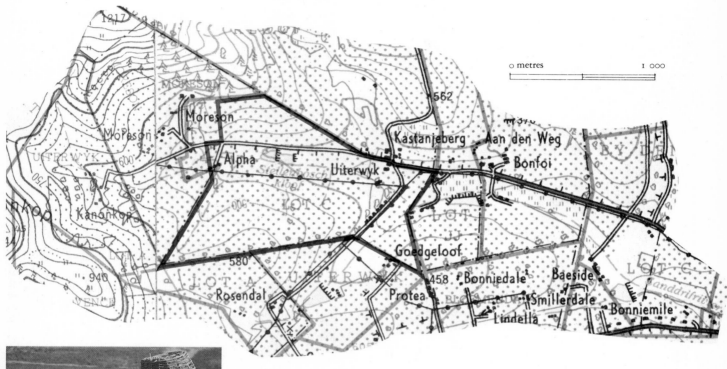

The Uiterwyk vineyards are on southern slopes and on the floor of the basin at the end of the Stellenbosch Kloof.

The name of this farm means 'outer ward'. In seventeenth century Holland the area surrounding the village of Kampen where the townspeople grazed their animals was called Uiterwyk. Today, Kampen has grown considerably and Uiterwyk is a suburb.

Dirk Coetzee, a son of Kampen, came to the Cape in 1679, the same year as Simon van der Stel. In 1682, Coetzee was one of the second group of settlers granted farms in the Stellenbosch area, and settled on the farm Coetzenburg, beside the eastern boundary of the future village of Stellenbosch. He was officially granted Coetzenburg in 1693 and received the unusual privilege of a second freehold grant in 1699, when he was declared the official owner of a piece of grazing ground, seven kilometres outside the village, which he named Uiterwyk. It is possible that Coetzee had been in possession of this land since 1682, as stated by Dorothea Fairbridge, and the title deed shows two buildings.

Johannes Krige evidently prospered at Uiterwyk, enlarging the house in 1791, building a large cellar in 1798, erecting substantial stables in 1812 and completing a second house in 1822. Willem Krige is recorded as having 75 000 vines in 1805 and 100 000 vines in 1824. The cellar built by his father was sufficiently large to be able to house Uiterwyk's produce until 1980, when Danie de Waal, the Estate's present owner, built a new fermentation cellar.

The De Waal family came to Uiterwyk from farms closer to the original settlement on Table Bay, including farms in Wale (De Waal) Street in today's Cape Town, Alphen, near Constantia, and Langverwacht, near Kuils River. Pieter 'Boy' de Waal bought Uiterwyk from the Krige family in 1864 and the farm has remained in the family ever since.

When Danie de Waal started working in the Uiterwyk vineyards for his father in 1938, only two varieties were being grown for the production of red wine. The wines made from their Cabernet Sauvignon and Shiraz vines were sold to a Stellenbosch wholesaler. Cinsaut was used to make white wine. White varieties were also grown, but the produce of most of these vineyards was sold as grapes. Danie took over a share of Uiterwyk on his father's retirement in 1946, and subsequently removed the Shiraz vines. The Cabernet Sauvignon vineyard was inherited by Danie's brother, Koffie, but was later bought back by Danie when he reconsolidated the farm in 1972. Uiterwyk lies in the south-western end of the Polkadraai, with most of the 110 hectares of vineyards on southern and eastern-facing slopes. All Uiterwyk's vineyards are irrigated. The soils have sections of decomposed granite and others of sandy loam, together with various degrees of mixtures. The basin, of which Uiterwyk forms the northern end, is sheltered from the south-east wind

Built by Johannes Krige in 1798, the old Estate cellar is now used to house red wines maturing in oak.

Built in the gabled style in 1791, Uiterwyk has been the home of the De Waal family since 1864.

and summer temperatures are a little warmer than those of other parts of Stellenbosch. The frosts that are found in winter in the vineyards along the Eerste River are unknown here.

Although Uiterwyk is best known for Cabernet Sauvignon, the white varieties outnumber the reds in the proportion 90 to 10. Uiterwyk's best known white wines are made from Colombard and Cape Riesling. The Colombard vineyards are planted in some of the richest soil on the farm, a philosophy that is shared by the neighbouring Spier Estate, where the rows run downhill, from west to east, on the east-facing slope. Danie believes this allows him to ripen the Colombard further and to harvest with a reduction in the normally high acidity of this grape, making a fuller, better balanced wine. He follows a similar practice with the Cape Riesling. The white grapes are lightly pressed to obtain extra flavour from the skins, with the pressed juice being fermented together with the free-run juice. The musts are fermented cold to give them greater delicacy.

The Uiterwyk vineyards provide Cabernet Sauvignon, Pinotage and Merlot grapes for the cellar to use. The Cabernet Sauvignon is grown in sandy loam soils, with rows running east to west to obtain an even rate of ripening for the whole crop. The grapes are normally picked at mid-ripe stage and fermented at cool temperatures down to about 12° Balling. The skins are pressed to obtain all the freely available liquid, and this is added to the fermenting must. Most of Danie's red musts are fermented in open stainless steel tanks, but some are fermented in the press, which is turned every two or three hours to mix the liquid and skins thoroughly. When the Cabernet Sauvignon has fermented dry, the wine is cleaned and then placed in large oak casks to mature. Most of the Pinotage is made as white wine. Danie makes a small batch, using a similar technique to that used for the red wine, for sale as Estate wine. The Merlot vineyards are young, but as they mature full-bodied wines will be produced for blending with the Cabernet Sauvignon.

Danie de Waal is the Uiterwyk vintner. His son Chris is in charge of the Estate cellar.

As Danie sells the bulk of his wines to a wholesaler, only a small quantity of Estate bottled wine is made available to the public.

Wine is for sale on the Estate.

Uitkyk

"I believe in lighter styles of red wines," says Harvey Illing of Uitkyk. "I have great regard for Cabernet Sauvignon. It produces many of the best red wines in the world, but for some reason South Africans tend to make rich, deep-coloured highly flavoured reds from Cabernet. Some wine makers then blend it with lighter reds made from other grapes to make the Cabernet pleasant to drink and therefore marketable at a younger age. Cabernet doesn't have to be a heavy wine. It's simply a matter of technique. Red wines get their colour and flavour from the skins of the berries. The longer you leave the fermenting wine on the skins, the richer the colour and flavour you'll get.

"I don't allow my Cabernet Sauvignon to ferment for more than three days on the skins. I then draw it off, settle it, and allow it to ferment dry by itself. This produces a lighter wine. It has a different colour, fragrance and flavour and you can drink it with pleasure at three years old."

A man named Martin Melck was employed by the Dutch East India Company in the years around 1742 to 'keep the natives quiet' wherever the Company had operations. Martin Melck was a German mercenary soldier who proved so highly efficient at his job that his name has become legendary in the annals of the areas where the Company held power. When Martin Melck tired of fighting, he married a widow who owned a wine farm in Muldersvlei, north of Stellenbosch. The farm was called Elsenburg, and is now an agricultural college owned by the South African Department of Agriculture.

It is not known whether Martin Melck had a son, but when his daughter married a man called Beyers in 1776 he ceded the farm that is now Uitkyk to them as a wedding present. While Mr. and Mrs. Beyers owned the farm, the French architect Thibault, who had been brought to South Africa and commissioned to design houses for several wealthy residents, built a town house on Uitkyk, and this is believed to be the only one of its kind on a farm in South Africa. Thibault was famous for his Georgian style of architecture, which contrasts strongly with the design of the standard Cape Dutch house. Uitkyk's front door is a study in wood sculpture. It was carved by the German sculptor Anton Anreith, who was also responsible for the frieze on the Groot Constantia wine cellar.

There were four Beyers generations on Uitkyk, and then in the 1880s the farm was sold to the Van Niekerk family, who used it principally for grazing cattle, until 1903, when they sold it to Senator J. H. Sauer. It is likely that the first vines to be seriously cultivated on Uitkyk date from this period. However, from its establishment until 1930, when Sena-

Uitkyk's Georgian-style mansion was built by Johan Beyers in 1788.

tor Sauer's widow sold the higher and larger portion of the farm, Uitkyk's main function had been to provide a stately country house for wealthy people.

In 1929 a Saxon nobleman, George von Carlowitz, decided that South Africa promised a better future for himself and his two sons than he could foresee in Germany, and they migrated. Uitkyk was on the market, and he believed that the well-drained slopes of the farm promised to make good wine-growing country, and that the steeper hillsides offered prospects for forestry.

Von Carlowitz organised the farm so that the sons ran the farm as a part-

Though Uitkyk is a high farm, the west-facing vineyards are able to fully ripen grapes.

nership. The eldest son, Hans, developed the wheatfields on the lower slopes and planted conifers on the steeper hillsides for timber production. The younger son, George junior, was responsible for the vineyards and wine production.

In 1939 Hans von Carlowitz returned to Germany, and George junior was left in control of all aspects of the farm. He built a large cellar, designed primarily for the production of red wine, as he made no provision for cooling equipment.

George von Carlowitz was interested only in the production of noble quality wines and planted vineyards of Cabernet Sauvignon, Cinsaut, Steen and Cape Riesling. He planted Cinsaut to blend with Cabernet, and made a wine that he named Carlonet. He planted Steen to add quantity to his Riesling production, and from these two varieties made a wine that he called Carlsheim, sweetish and fruity.

George never had more than 35 hectares under vine at one time. He and his wife Hilda supervised the wine-making process and were sufficiently skilful to ensure that Carlonet and Carlsheim took pride of place in the cellars of the small number of people who drank quality table wine in the Cape Town of the 1940s and '50s.

During the 1950s George was well aware of the need to introduce cooling equipment in his cellar to improve his Carlsheim, and to replace old vines on the farm. Unfortunately he did not have the money and, as his son was not interested in wine farming, he was forced to sell.

The buyer was Gerry Bouwer, who had interests in mining and a passion for rebuilding Uitkyk as a wine Estate. His business commitments prevented any thought of his living permanently on the farm and overseeing the redevelopment necessary. His daughter had married a dentist named Harvey Illing, who practised in Durban, and Gerry suggested to Harvey that he give up dentistry for viticulture. That was in 1963, the year of the purchase, and Harvey Illing has lived on Uitkyk ever since.

The first job of the new team, Bouwer and Illing, was to appraise what was left on the Estate that could be salvaged, and to decide what would need to be replaced or extended. As George von Carlowitz had always hoped to extend his wine production, he had built a cellar that was at least twice as big as his needs. The new owners decided to leave the cellar untouched and give first attention to the vineyards, where they had 35 hectares of old vines, some of which had begun to decline.

During the next ten years extensive areas of bush were bulldozed to make way for new vineyards. Because of the mountainside location of the farm these new slopes faced east, southeast, south, south-west and west. Locations were chosen for their suitability of soil type and the direction in which they faced, this being important because, situated high on the slopes of Simonsberg, Uitkyk is prone to easterly gales, and certain types of vines are better equipped than others to survive and produce under these conditions.

In addition to the existing 35 hectares under vine, another 120 hectares were cleared and planted during those ten years. Half of this area was planted with Cabernet Sauvignon, about one quarter with Steen, and the rest evenly divided between Clairette blanche, Cape Riesling and Cinsaut.

Then attention was turned to the old vineyards, where vines found to be past their best were taken out and replaced with Shiraz and Pinotage.

From 1963 to 1968 Harvey Illing attempted to continue Von Carlowitz's tradition and supply Carlonet and Carlsheim to long-established customers, with a view to pushing up sales of each as new areas of vines came into production. There were two main obstacles to this programme. The first was the amount of time Harvey was spending on the deliveries. Demand had dwindled and, when orders came in, he would take several cases to the railway station and

Harvey Illing, who gave up a dentist's chair to assist his father-in-law on historic Uitkyk.

forward them to Bloemfontein or Johannesburg. And once a month he would take a van with a few cases into Cape Town and deliver them from house to house. There was far more nuisance than profit involved, and not even retention of goodwill was certain. The second obstacle was that, though George had built a big cellar, the equipment inside it was outmoded, with a lack of cooling equipment for the improved production of white wine.

The year 1968 was the first of the seven years that Uitkyk spent off the Estate market. Modern equipment for the processing of both white and red wines was installed in the cellar. In 1969, when modernisation was completed, Uitkyk continued to make red wine, but sold it together with the Estate's production of white grapes to a wine merchant. The next major development was Uitkyk's association with Die Bergkelder, which guaranteed national market distribution.

Like George von Carlowitz, Harvey Illing and Gerry Bouwer saw the many variations of slope and elevation on Uitkyk as the source of both red and white wines. They planted a small vineyard of Shiraz on the lower slopes near the cellar, and extensive vineyards of Cabernet Sauvignon, chiefly on the most elevated sites available, to get the best use of the granitic, mountain soils, as well as Cape Riesling and Clairette blanche on the warmer soils towards the centre of the farm. Of all the vineyards planted by Von Carlowitz, only the Steen was retained in the redevelopment programme. The first of the new generation of Uitkyk wines appeared when two of the 1973 red wines, a Cabernet Sauvignon (called Carlonet) and a Shiraz, aged in oak, were released together with a dry Carlsheim, a Riesling and a semi-sweet Late Harvest.

The small quantity of Shiraz had been so promising, even in youth, that Harvey had planted a larger Shiraz vineyard in 1974 to swell the production for the future. Each wine Estate has unique conditions, and Uitkyk was to teach Bouwer and Illing that a situation 500 metres above sea level, on the exposed windy slopes of the Simonsberg, provides too cool a micro-climate for Cabernet Sauvignon to ripen fully a reasonable crop of grapes. In time, after the lessons of several harvests, during cool and warm years, the team planted another Cabernet Sauvignon vineyard on the lower, warmer slopes and removed all the red vines planted at the top of the farm.

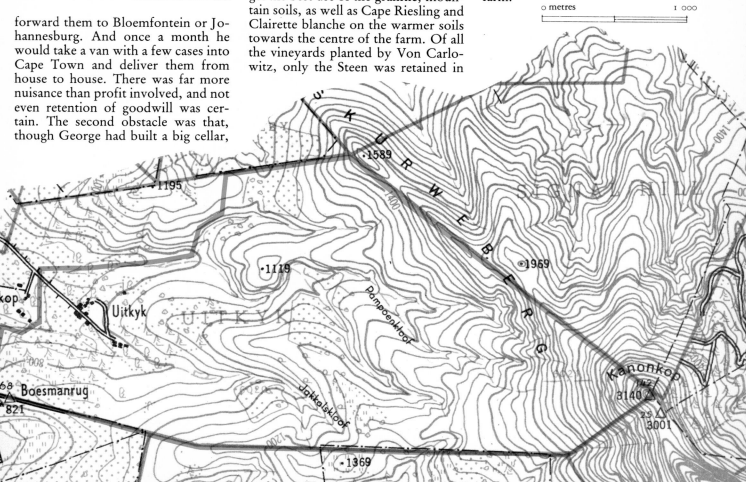

Harvey has found that Uitkyk's 300-metre variation in altitude and its varied slopes do indeed provide ideal growing and ripening conditions for red and white varieties, but, contrary to what he first imagined, the cooler positions are preferred by the early-ripening whites, with the mid- and late-season reds benefiting from the warmth on the lower part of the farm. Accordingly the early varieties Chardonnay and Pinot gris and the mid-season's Sauvignon blanc have been chosen for the most elevated and exposed vineyards. The Sauvignon Blanc will be blended with selected dry Steen to produce the fruity-flavoured Carlsheim, while Chardonnay and Pinot Gris, in time, will probably be marketed as pure varietal wines. The Carlonet now produced by the lower vineyards will be marketed alongside Shiraz, which has been revived as a cultivar wine.

The red musts are fermented at cool temperatures (about 20 °C) in closed rotor tanks, where the mixture of skins and juice is agitated every three hours, until sufficient colour has been extracted, when the husks are removed. After the completion of fermentation the red wines are moved to Stellenbosch for maturation in oak.

The grapes for the dry white wines are picked mid-ripe (21-22 ° Balling) and, after crushing, the juice is left with the broken skins for 8-12 hours, for maximum flavour. This period of skin contact is not used in the production of Cape Riesling (which can turn brown) or Pinot Gris (which will turn golden yellow). The grapes for the Late Harvest are allowed to ripen further before picking, and are allowed to have skin contact. The white musts are fermented cold (12-14 °C) until either dry or, in the case of Late Harvest, until they have reached the required degree of sweetness, and then racked and cleaned, ready for delivery to the Stellenbosch cellars of the Bergkelder for bottling. Uitkyk white wines are not matured in oak.

Wine is not for sale on the Estate.

After crushing, Uikyk's Cape Riesling juice is drained off and not given a period of skin contact.

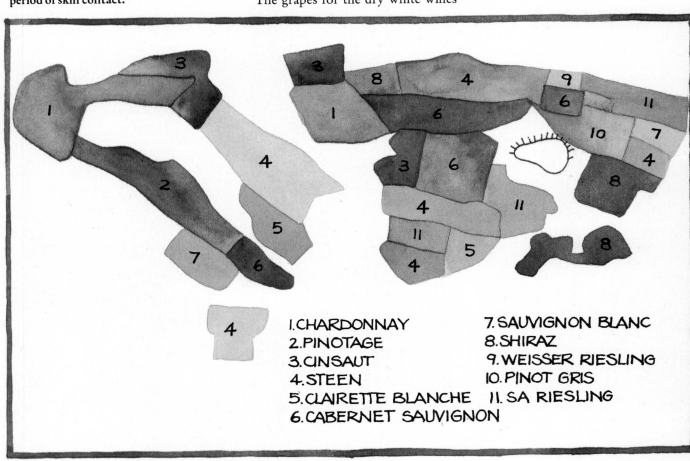

1. CHARDONNAY
2. PINOTAGE
3. CINSAUT
4. STEEN
5. CLAIRETTE BLANCHE
6. CABERNET SAUVIGNON
7. SAUVIGNON BLANC
8. SHIRAZ
9. WEISSER RIESLING
10. PINOT GRIS
11. SA RIESLING

Verdun

Verdun's vineyards stretch across the rolling granite hills called Bottelary that lie between False Bay to the south and Simonsberg to the north. This land is now important and valuable wine country, but until 1916 it was considered fit for little else than grain farming.

For two centuries before them the farmers in the Stellenbosch area believed that good wines were to be grown only on the flat lands to the south and east, by the river. Professor C. J. Theron, a famous figure in South African wine history, is said to have stated that the best red wines of South Africa were to be found on the farms on the foothills and lower slopes of the range of mountains to the south-east, north and north-west of Stellenbosch.

Kobie Roux's grandfather bought the property that is now Verdun, and also the land on the other side of the main road, stretching down to the river. On this latter land he built his home, planted his vineyards and ran his business. In 1915 Kobie's father said that with a little luck the Oubaas might let him have the 'sonder naam' farm across the road. Kobie went to university in Stellenbosch to learn the technical details of farming, and viticulture in particular. He was too young to join the South African forces that were fighting in France at the time, and his grandfather duly gave him the hillside farm, to do with as he pleased.

When Kobie found that indeed he had inherited the 'sonder naam' farm on the hills away from the river, the first problem was, obviously, to give the farm a name. A battle between German and combined British, Colonial and French troops was being fought in a long and bitter struggle on the fields of Verdun in France, at the time Kobie Roux took possession of the farm. His nephew suggested that he name the farm Verdun, for he was sure to have as great a battle with the land as the South African soldiers were having with the German army.

Kobie immediately planted vines, most of which were red wine varieties, and was soon harvesting and making wine. The next step was to plant some of the better-quality cultivars, and Kobie was probably the first wine farmer to plant Gamay and Cabernet Sauvignon vines far from the rich soils by the river and mountain range.

Kobie was a traditional grape grower and wine maker. The majority of the vines on his farm were untrellised and unirrigated. He grew Steen, Riesling and Raisin blanc for the making of white wines, and Gamay, Cabernet Sauvignon, Cinsaut and Pinotage for his red wines. The vines were pruned back to prevent overbearing and the grapes were picked full-ripe to make full-bodied wines. He fermented all of his musts in open concrete tanks and matured the reds in large oak casks for at least two years. Until 1971, all of these wines were sold to a wine merchant in Stellen-

Though most of the Cinsaut vineyards have been removed, a dry red is still made.

Winter cuttings from selected Gamay vines will be used to establish a new vineyard of this variety.

Like most wine Estates, Verdun uses the spring season for bottling both red and white wines.

Verdun is situated on the southern slopes of the Bottelary Hills, on the western edge of Devon Valley.

bosch. In 1970, Kobie decided to bottle a small quantity of Cinsaut and Gamay for sale to visitors to the farm and he became a founder member of the Stellenbosch Wine Route.

Kosie, Kobie's son, joined his father on the farm in 1951 and began to influence the course of Verdun in new directions. He planted Sauvignon blanc, Weisser Riesling, Gewürztraminer and Kerner to swell the choice of white wines and added Pinot noir to the list of reds. The first Estate bottlings of Sauvignon Blanc, Weisser Riesling, Gewürztraminer and Pinot Noir demonstrated to Kosie that the public wanted these wines. "I find the quality white varieties more difficult to handle than Steen or Riesling," Kosie says. "It's harder to pick the grapes at the correct stage of ripeness, because it takes longer to pick the same quantity. I get one third more each day if the pickers are working in one of the old vineyards."

Kobie, who died in 1976, was by preference a producer of red wines, but Kosie has changed the approach. Many of the Estate's old red vineyards have been removed and replaced by blocks of white cultivars. The proportion of white to red in the 100 hectares of vineyards is now 70 to 30, and this is likely to remain the pattern for at least the next few years. Kosie removed almost all the old Cinsaut vineyards and has only replaced one with young Cinsaut vines, which he feels are a disappointment. The crop is larger but the wines lack quality.

He has replaced most of the Gamay vines planted by his father with new vineyards of Gamay, but in this case believes that the quality of the younger vines is superior to the older ones. The Gamay produces about seven tonnes to the hectare.

The Verdun Estate has northern slopes, eastern slopes, southern slopes and a depressed shallow valley in between. Kosie has found that the vines grown on the north-facing slopes ripen their grapes earlier, and he believes that they produce some of his best wines. His Steens produced from grapes grown in these vineyards are often better wines than those made from later-ripening blocks. His Weisser Riesling vineyard is also on this slope and has produced a consistently fine-quality wine. Verdun's Cabernet Sauvignon vineyard is planted in the slightly more fertile soil in the depression between the slopes and produces a medium-bodied wine.

Another change to the traditional Stellenbosch style of vine growing was Kosie's implementation of irrigation on about half of the Estate vineyards. He has installed permanent lines of micro-sprays and hopes to be able to prevent damage to both quantity and quality of his crop in prolonged dry spells.

The Estate's cellar and the styles of wine have also undergone modification. To produce the lighter, fruitier

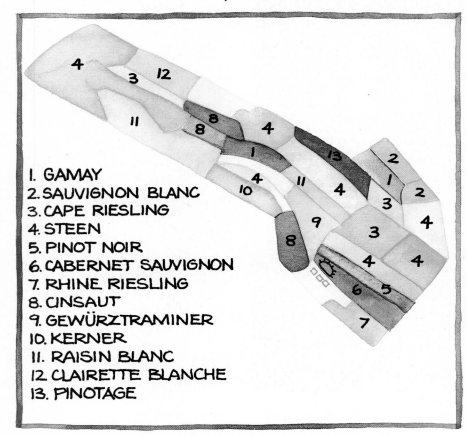

1. GAMAY
2. SAUVIGNON BLANC
3. CAPE RIESLING
4. STEEN
5. PINOT NOIR
6. CABERNET SAUVIGNON
7. RHINE RIESLING
8. CINSAUT
9. GEWÜRZTRAMINER
10. KERNER
11. RAISIN BLANC
12. CLAIRETTE BLANCHE
13. PINOTAGE

Harvesting late in the season for full-ripe grapes to obtain extra body for blending wines.

whites and reds that are gaining popularity, Kosie picks his grapes with a little less sugar than was his father's custom. The cellar has seen the addition of closed fermentation for the white wines. The chilled juice is inoculated with yeast and fermented at 15–16 °C until the sugar has either been fermented out or the must has reached the degree of sweetness that Kosie prefers. Fermentation is arrested by chilling the must and then filtering the wine. The red grapes are normally picked at around 22 ° Balling, and after crushing the skins are mixed with the must fermenting at a temperature of about 16 °C until the sugar content of the must reaches about 16 ° Balling, at which point the skins are removed. The must is allowed to ferment at 16 °C until all of

Verdun faces south-east towards the shallow valley of the Eerste River and the imposing peaks of the Helderberg.

Kosie Roux, Verdun's winemaker, has the Cape's only commercial vineyard of Gamay.

the sugar is converted and the wine is dry. The new wine is then racked and prepared for wood maturation. Pinot Noir normally receives twelve months wood ageing before bottling. The Gamay and Cabernet Sauvignon stay in the large casks for up to two years. These wines are only given a short period of bottle maturation before release to the market.

Wine is sold on the Estate.

Vergenoegd

South African wines have regularly won prizes and medals at wine shows throughout the world. Estates covered in these chapters have won medals in all wine-producing countries, but the legislation governing the registration and operation of Estates in South Africa forbids the mention of these prizes on Estate labels. Consequently little mention has been made of the honours won by individual Estates.

For Vergenoegd, it is worth making an exception. From 1971 to 1974, the Faure brothers of Vergenoegd won the trophy for the best wine of the show at the annual Cape Wine Show with Cabernet Sauvignon from their Estate. Four successive vintages, four different Cabernet Sauvignon wines, four Grand Champions. There is no longer a class for the best wine of the show, so the record may never be beaten.

The first settler on the farm Vergenoegd was Pieter de Vos, who began to clear the ground by the side of the Eerste River, near False Bay, in 1692. His farm was the last along the river

Exposed to summer winds, just across the sand dunes from False Bay, vines struggle but produce quality grapes on Vergenoegd.

and adjoined land settled by Henning Hüsing and Ferdinand Appel. De Vos was granted the farm by Simon van der Stel in 1696, but he died soon after. His widow sold the farm to her neighbour Appel in 1700, and the property was handed down by direct descent until the first Faure bought the farm in 1820. Johannes Gysbertus Faure transferred the farm to his brother, and the farm has been owned by succeeding generations of Faures ever since.

Ferdinand Appel was one of the first Cape-born burghers to take up a farming life, and became involved, with his neighbour Hüsing, with the faction that circulated a petition to have Willem Adriaan van der Stel censured for corrupt use of his position as Governor. To quell the growing revolt, Van der Stel had five of the ringleaders arrested, banned from the Colony and deported to Holland. Appel and Hüsing were among the

Jac Faure, who with his brother Brand runs the Vergenoegd Estate.

five who managed to smuggle an erudite letter, penned by Hüsing's nephew Adam Tas, describing the misuse of Company funds by officials, back to Holland. They were thus instrumental in removing the Governor from his position—after which Appel and Hüsing returned to their farms. The Vergenoegd manor house is the work of Johannes Colyne, who in 1773 enlarged and improved the existing homestead. The stables, cellar and other historic buildings were added over the next hundred years.

It is believed that wine has been made on Vergenoegd since 1700, but it was not until the formation of the KWV that wine became the major income-earner for the farm. The first event of significance in Vergenoegd's wine history was its association with

Stellenbosch

The Vergenoegd house is a classic example of Cape Dutch styling and has been the Faure family home since 1820.

Vineyards planted on the Estate's deep-drained alluvial soils require additional water during summer.

After maturation in large oak casks Vergenoegd's red wines are filtered before bottling. The majority of the wines are sold to the KWV.

Charles Niehaus, then senior lecturer in viticulture and oenology at Stellenbosch University. Dr. Niehaus was a microbiologist who had set himself the task of introducing the 'flor' yeast used in sherry-making to South Africa. He had brought several samples of Spanish 'flor' yeast cultures back from Europe after completing a three-year study course in Germany, and during 1935 he was probably the first man outside Spain to get 'flor' yeast to develop on a wine. This work was done in a laboratory in the University of Stellenbosch.

His success came to the attention of Dr. Perold, who was at that time chief wine and brandy expert at the KWV. Dr. Niehaus believed that, as the climate and geographical position of South Africa's Cape Province were very similar to those of Spain, the home of sherry, native 'flor' yeast must exist in the southern continent. In 1936 Stellenbosch University and the KWV agreed to co-operate to give Dr. Niehaus the chance to find 'flor' yeasts on South African wine farms. During that year, from samples of wine lees taken from South African farms, he was able to isolate eighteen different kinds of 'flor'.

One of the farms on which he found the yeast was Vergenoegd. In 1937 Dr. Niehaus joined the KWV as sherry expert and persuaded John Faure to make sherry. Vergenoegd has produced sherries from Steen grapes for the KWV ever since. It was not long before Dr. Niehaus became Dr. Perold's assistant in charge of good wine as well as sherry, and eventually he became the KWV's chief wine expert.

The KWV had bought a lot of Vergenoegd Shiraz over the years prior to Dr. Niehaus's association, and had made sweet fortified red wine from it. One day when Dr. Niehaus was visiting Vergenoegd, John Faure showed him an old Shiraz vineyard which he intended to take out, and asked Dr. Niehaus's advice on which new variety to plant. His advice was Cabernet Sauvignon.

"I told him," says Dr. Niehaus, "'you have the right soil, the right climatic conditions and you're close to the sea. You produce the Cabernet and I'll buy it.'

"The next year when I called at the farm, I saw the young vineyard. I said, 'Jan, did you plant that vineyard to Cabernet?'

"He said, 'No, man, I was just a little bit afraid. I planted Shiraz. I know what Shiraz can do on this farm.'

"'Damn it!' I said to him, 'Didn't I tell you to plant Cabernet?'

"'Well,' he said, 'the other part of that vineyard is coming out and I promise I'll plant Cabernet.' And he did. That was the origin of Cabernet on Vergenoegd. It is from that vineyard they produced the Cabernet Sauvignon that won the Grand Champion wine trophy of the Cape Show four times in a row. I only wish that Jan Faure could have lived to see the success that his sons Jac and Brand are having with that Cabernet."

On level ground beside the river, with an average elevation of about twelve metres above sea level, Vergenoegd has two major types of soil that can be used for agriculture. The deep drained alluvial soil beside the river is too fertile to be a base for vineyards producing high-quality wine. The vines grow very well, but are unable fully to ripen their grapes. Adjacent to the rich soils are sandy soils with a layer of clay not far beneath the surface. All Vergenoegd's best wines are made from vineyards grown in the sandy soil, with roots penetrating into the clay. The clay provides the moisture required especially by the late-ripening Cabernet Sauvignon to bring its grapes to maturity toward the end of summer.

The life of a vine on the Vergenoegd Estate is not easy. The vineyards are planted on poor soils and receive little rainfall each year. The farm is less than five kilometres from the sand dunes of False Bay and receives south-east and southerly winds throughout spring and summer. The vines are given one burst of spray irrigation each summer to prevent premature ripening owing to drought conditions. All the vines are trellised on a low system to protect sensitive

Vergenoegd

This historic picture of harvesting on Vergenoegd was taken in the time of the Faure brothers' grandfather, possibly seventy-five years ago.

varieties from wind damage. Nevertheless, in a difficult ripening year, Jac Faure picks his Cabernet Sauvignon at 21° Balling, and in warmer years the grapes arrive at the cellar with up to 24° Balling sugar content. The must, with skins and juice together, is fermented with dry yeast in open concrete tanks until about half the sugar has been removed. After removing the skins, Jac ferments the liquid in the open tanks, where he is able to lower the temperature of the must until approaching dryness, and then the must is placed in closed containers, occasionally using 4 000-litre maturation casks for this purpose. Once dry, the wine is allowed to have a malolactic fermentation before going into wood for as long as three years.

Only red wines are made in the Estate cellar. The Estate's white grapes are sold to the KWV. Only limited quantities of Cabernet Sauvignon, Shiraz, Pinotage, Tinta Barocca and Cinsaut are sold to the public.

Wine is for sale on the Estate.

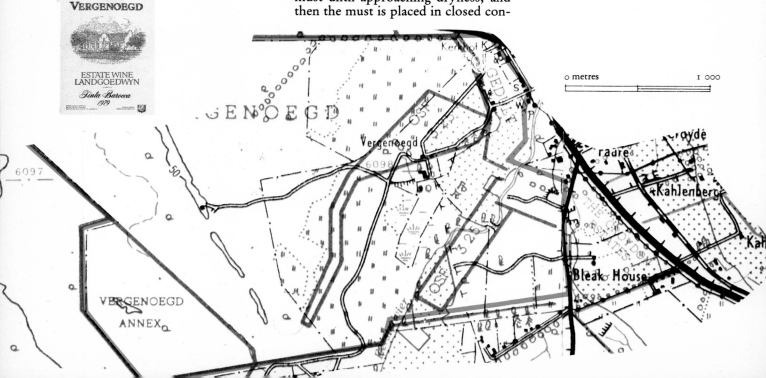

Vriesenhof

The elevated southern slopes of Vriesenhof provide just sufficient light and warmth to ripen the remaining Cabernet Sauvignon vineyard.

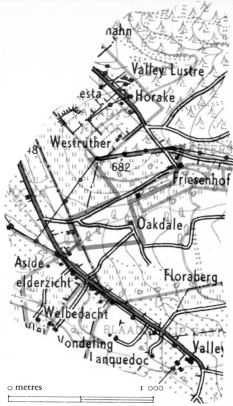

1. CHARDONNAY
2. CABERNET SAUVIGNON
3. PINOT NOIR

Looking south-east across Vriesenhof towards the forests at the foot of the Helderberg Mountain.

It was during his viticultural student days that Jan Boland Coetzee was offered the opportunity by Paul Sauer to manage Kanonkop, one of the country's most highly regarded red wine farms. Jan became a part owner, and together he and the Estate built reputations for solid, reliable quality.

In 1980 he decided to strike out on his own and resigned from the Kanonkop Estate. He started looking for a property small enough for him to be able to handle on his own, with a price tag within his reach, yet with the potential to produce high quality wines. He found an eleven-hectare piece of vineyard land on an elevated site on the slopes of Stellenbosch Mountain. The farm, Vriesenhof, had an old house with a production cellar built into one end of it, a not uncommon practice on small properties. He also negotiated a long-term lease on an adjoining ten hectares of land with identical soil and aspect, sharing a boundary with the most eastern Blaauwklippen vineyards.

Most of the 21 hectares have a fairly constant southern slope, with about three hectares on the crest of the slope, almost level. The vineyards are the most elevated in the area and, facing south on the rise and crest of a substantial ridge, have an active flow of ventilating wind in summer. The eleven hectares of freehold were already planted with Cabernet Sauvignon and Pinotage at the time of the purchase, while the leased land was covered with bush.

Vriesenhof was originally part of Kafferkuil, a rented farm dating back to the days of Willem Adriaan van der Stel, when it was used for hunting and grazing. Sir John Cradock changed the rent farm system to perpetual quitrent in 1813, with Kafferkuil being one of the first grants to be made.

The small Estate has rich Hutton soils of decomposed granite with a base of clay not far beneath to hold moisture through the growing period. The soils are moderately fertile and the vines grow well. Jan has found the growing conditions in Vriesenhof's elevated vineyards to be so cool that the lower section of the Cabernet Sauvignon on the steepest slope did not fully ripen its grapes even in the warmth of 1982. Jan has replaced these vines with Pinot noir.

Wine had not been made in the small and basic cellar since 1946, and for his 1981 maiden vintage Jan had to purchase a crusher, some red wine fermentation tanks and some large wooden vats. The season had been cooler than normal and rain fell in uncomfortable quantities during the ripening period of Cabernet Sauvignon. Late in April Jan picked his grapes at a relatively low sugar con-

Jan Boland Coetzee and his small team are able to harvest the farm's 7 hectares of Cabernet Sauvignon within a few days.

Jan found that the coolest part of his Cabernet Sauvignon vineyard, high on the slopes of the Stellenboschberg, was unable to ripen properly.

tent, but found his must to have plenty of acidity and a low pH. The grapes were unlikely to have ripened further, and the soil and the variety retained the vitality of the grape to a rather late date.

In July of that year Jan and his family left for France to realise his ambition of gaining experience in the French wine industry. Jan worked as a cellar assistant for a merchant in Beaune, Burgundy, buying, blending and bottling Pinot Noir and Chardonnay wines. He returned to Vriesenhof in time to prepare for the 1982 harvest, which was a beautiful ripening season, with plenty of sunlight. Jan picked his crop in two batches, four days apart. The first saw must with an average Balling reading of 21,5 ° and an acidity of 10,4 gℓ arrive in the cellar. The second batch of musts averaged 22 ° Balling, but the acidity had fallen to 7,5 gℓ, making Jan realise that the acid values of his new vineyards were indeed fallible and that his young vines had a fairly critical harvesting point.

Jan ferments his Cabernet Sauvignon musts in open stainless steel tanks at between 22 and 25 °C, and varies the period that the skins are left in the mash according to the sugar concentration of his crop. In 1981 the wine

Vriesenhof commenced production with fermentation tanks, cooled by chilled water, outside the cellar building.

finished with only 10,5 ° of alcohol, and he kept the skins in the must until it was almost dry. In 1982, with ideally ripe grapes, he removed the skins at a much earlier stage. After the fermentation has completed its task, he takes the new wine off the yeast and allows it to settle and, it is hoped, begin a malolactic fermentation. South African wine makers are permitted to add acid to grape juice or wine to improve its balance. Malolactic fermentation has the tendency to remove much of this added acid. At Vriesenhof Jan has worked with grapes with high natural acidity, which after both alcoholic and malo-lactic fermentations remains present in the wine. However, being a natural component, it is hardly obvious. In youth, these young wines with high acidity are soft and fresh.

At the time of writing, Jan makes only a pure Cabernet Sauvignon wine from the Vriesenhof vineyards. His plans are to produce only three wines with this label: a Cabernet Sauvignon–Merlot–Cabernet Franc blend, Chardonnay and a small quantity of Pinot Noir. The additional red vineyards were planted in 1982, and Chardonnay planned for the following year.

Jan's small cellar, built into the old house, will be replaced by a larger pressing and maturation cellar before the crops from the added red vineyards are ready to harvest. Because he still buys some grapes each season to add to his small crop, Vriesenhof is not yet a registered wine Estate.

Wine is not yet for sale on the farm, but succeeding vintages will be available from the tasting room on the property at certain times during the year.

Zevenwacht

The broad expanse of Zevenwacht's vineyards slopes south towards False Bay and receives cooling breezes in summer.

The Zevenwacht manor house, recently restored, was built in 1800.

High on the slopes of the Bottelary Hills, Zevenwacht lies near the town of Kuils River and faces Cape Town across the broad level expanse of the Cape flats. The farm, formed by the amalgamated Zevenfontein and Langverwacht properties, contains 354 hectares of land, the majority of which run along the south- and south-west-facing slopes of the granite-origin hills. The ends of the farm tuck over crests to the north, facing Durbanville and the Atlantic, and the east, into the Stellenbosch Kloof, facing the Helderberg Mountain. The larger portion of the farm is angled toward False Bay, and being elevated gets the southerly breezes that come off the water during summer. There are more than 150 hectares of high ground with a southerly aspect, planned for vineyards.

The Zevenfontein enterprise is operated by a public company, owned by several hundred wine-loving shareholders, and is headed by one of South Africa's leading architects, Gilbert Colyn.

In 1973 Gilbert bought a 39 hectare farm, Avonduur, on the slopes of Banhoek near Stellenbosch to provide a country retreat where he could indulge his farming ambitions. In conjunction with Rudolf Scholms, the farm's manager, Gilbert planted Cabernet Sauvignon, Weisser Riesling and Tinta Barocca vines, but the scale of the project was too small, and in 1978 Gilbert bought Zevenfontein with 200 hectares of land available for vineyards. The property was partially stocked with standard quality varieties and was supplying grapes to a co-operative cellar. Though the property had an historic house, the accompanying cellar had vanished without a trace. Two years after buying Zevenfontein, Colyn was able to add the adjoining Langverwacht property to form a combined unit of considerable size.

The consolidation of Langverwacht and Zevenfontein was apt, as the two were in the past operated as one property. Langverwacht was granted to Jean le Roux de Normandie, who arrived at the Cape in 1688. He was granted a farm in the Dal Josaphat area near Paarl in 1692. Langverwacht was his second farm. The part of Langverwacht that dips over into Stellenbosch Kloof was probably the original freehold land, as Langverwacht was described in this way in early documents. In 1793 an area of adjoining land, known as Zevenfontein and traditionally used by the owners of Langverwacht as grazing land, was granted to Daniel Bosman, owner of Langverwacht, and the whole passed to Petrus Hiebner in 1799. Hiebner built a manor house in the following year. In 1825 the prop-

Zevenwacht has twice as many white variety vines as red and during the early years will be best known for white wines.

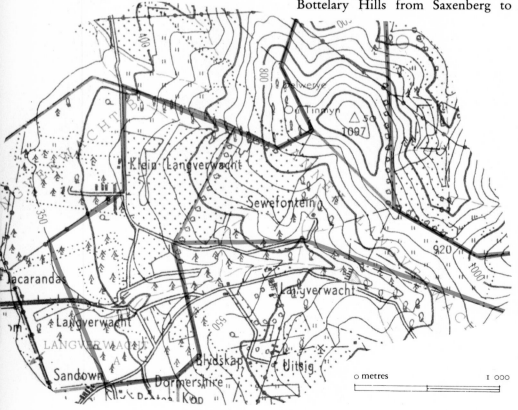

erties were owned by Jacob Malan, who had 55 000 vines and had fifteen leaguers of wine in his cellar. During the nineteenth century, geologists found traces of tin and tin-bearing ores among the granite-origin rocks and soils in the line of deposits running along the east-west axis of the Bottelary Hills from Saxenberg to Hazendal. The most promising indications were on Langverwacht, and the Good Hope Tin Mine extracted a few tonnes of ore each month during a period after the Second World War.

Rudolf Schloms joined Colyn in the Zevenwacht project and has embarked on a programme of soil analysis and vineyard preparation prior to planting to allow the chosen varieties to produce to their optimum potential. The Clovelly and Hutton soils that are found on the slopes have a higher calcium content than similar soils closer to Stellenbosch, and most of them do not require additional lime. Where necessary, the pH of the soil has been adjusted. Most of these soils on Zevenwacht contain a percentage of clay and are based on a layer of clay that ranges from one to two metres below the surface, retaining moisture from the winter for use by the vine roots in summer. The average rainfall on the property is 700 mm per year, enough to grow vines and ripen grapes during most years, but Gilbert Colyn's policy of leaving nothing to chance has resulted in the provision of irrigation in all vineyards. Additional water is only provided to assist the growth of very young vines.

The objective is to provide optimum growing and ripening seasons each year, so that though the vine is forced to struggle to provide for its crop, the degree of difficulty is limited to prevent damage to the vines or a drastically reduced harvest. When the two properties were reunited in 1980, Gilbert Colyn found he had vineyards of Clairette blanche, Palomino, Chenin blanc, Pinotage and Cinsaut, and began to plan the replanting schedule for the soils on Zevenfontein that were being prepared by Rudolf Schloms. In 1981 the team planted 30 hectares with Weisser Riesling, Gewürztraminer, Chenin blanc and Cape Riesling vines. The following year Sauvignon blanc and more Weisser Riesling vines were added to the white variety vineyards and Cabernet Sauvignon, Merlot, Cabernet franc, Shiraz and Pinot noir vineyards were planted for the production of red wines. These young vineyards will be allowed to come into production in the shadow of the established vineyards, and be allowed to attain some maturity before they are expected to carry the laurels of the cellar.

The choice of varieties has been made by Gilbert in collaboration with Neil Ellis, previously cellar master at Groot Constantia and now in charge of wine making at Zevenwacht. Together they have chosen to plant white varieties in 70 per cent of the available vineyard soil, and a 70/30 proportion of white to red vines is part of the forward planning.

For several years Gilbert was involved in cellar planning for the Groot Constantia Estate, which culminated in the construction of a major new cellar in 1982. Research into cellar design took Gilbert to cellars in California and Europe, and this experience was used to design a sophisticated, partly underground cellar at Zevenwacht, built under the direction of Neil and Gilbert in 1982.

At Groot Constantia Neil had several seasons' experience with the noble Pinot noir variety under South African conditions. He believes that this vigorously growing vine can make superlative wines in the Cape if allowed to ripen a moderate quantity of grapes, in a restricted manner. For this cultivar he chose a vineyard on

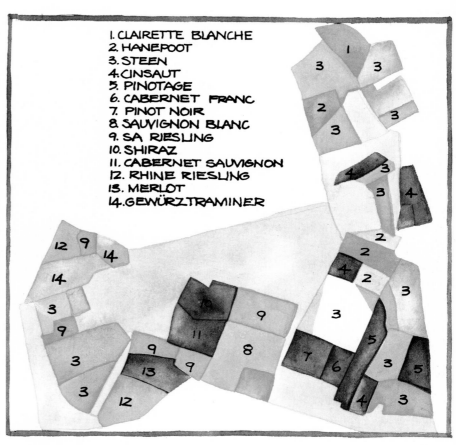

the lower slopes of the Langverwacht portion of the farm, where pebbly soils and a west-facing aspect provide the vines with warmth throughout the ripening season of this early variety. He planted a high concentration of vines within the block, without trellising and without irrigation. The objective is to encourage each vine to produce few bunches, little unnecessary wood or leaf cover and to provide each grape with a high concentration of nutrients.

The issue of public shares in Zevenwacht was calculated to provide the enterprise with sufficient cash reserves to develop large areas of vineyard and to build a capacious modern cellar. Since the shareholders now have a special interest in consuming and promoting a range of products in which they have a stake, they were offered the opportunity of obtaining special, limited-quantity wines and a private bottle-maturation cellar for storage or entertainment within the cellar building. In addition, Zevenwacht has substantial areas of land that are unlikely to be planted with vines, and shareholders will be invited to construct holiday cottages that make constructive use of the space and the views that are available. There are 70 hectares of land facing the Atlantic Ocean and several rocky outcrops facing Cape Town, False Bay and the Eerste River Valley that provide excellent sites for dwellings.

Zevenwacht is not a registered Estate, at present. However, this position may change in the near future.

From the 1983 season onwards wine will be for sale on the property.

Zevenwacht

SWARTLAND

208   *Swartland*

Allesverloren

During the eighteenth century this farm was an outpost of the European civilisation then developing in the Cape. The site of the farm had been chosen because of its elevated position on the side of a mountain named Kasteelberg. During the latter part of that century, a family farming the property used to make an ox-wagon trek into Stellenbosch every three weeks to attend church. After one such visit, they returned to find that a tribe of Bushmen had raided the farm, burnt down the buildings and scattered the cattle far and wide. The family must have stood and stared. There can have been only one thought in their minds. *Allesverloren.* All is lost. However, they were a pioneering family, and set to work to rebuild and re-establish the farm. Thanks to their perseverance, Allesverloren became and has remained a valuable and historic farming property.

The farm was originally granted by Willem Adriaan van der Stel in 1704 to a widow Cloete. The present owners, the Malan family, first arrived at Allesverloren in 1870, when Daniel François Malan came from Wellington and purchased this 700-hectare wheat farm on the slopes of Kasteelberg. There were some vineyards on the farm, but Malan was the first owner to take a personal interest in developing the vineyards. At first this project was meant to supply just the needs of the family, but Malan found a ready market in selling his fortified sweet wine to other wheat farmers.

The first son, also named Daniel François, was given a choice. Either he could finish his studies at school and return to help develop and eventually take over the farm, or he could continue to study and allow one of his brothers to take over. He decided to proceed to university and forfeit the chance to own the farm. He chose wisely, for his studies allowed him to embark upon a career that eventually led him to become a member of Parliament and subsequently Prime Minister of South Africa: Dr. D. F. Malan.

The second son, Stephanus François, took over the farm in 1904. At

Looking north-east from Allesverloren's most elevated vineyards, down on to the wheat fields on the valley floor.

Allesverloren

Looking to the north-east down the Allesverloren vineyards from just below the cliffs of Kasteelberg.

this time it had much more land under wheat than vine. Stephanus planted a number of new vineyards, but was much more interested in community affairs than in farming. His own eldest son, another Daniel François, was the first member of the family to become seriously interested in wine farming.

With the co-operation of Professor Theron and others on the staff of Stellenbosch University, Daniel began to make port-type wines on the farm. He obtained the materials for new red vineyards and developed them under the supervision of the university's viticultural faculty. He planted blocks of the well-known port varieties. Most of these are still in production today on Allesverloren. Over the years he was able to build up one of the most extensive collections of red wine cultivars in South Africa. He planted Tinta Barocca, Tinta Roriz, Souzão, Malvasia Rey, Shiraz, red Muscadel and some Cabernet.

Daniel became a port-wine maker of renown in South Africa, twice winning the trophy for the best wine on the Cape Show with ports.

Allesverloren passed to the fourth generation of Malans in 1961 when the farm was again divided between two sons. Fanie and his brother Gerard bought the Estate from their father. Fanie, the elder, received the house and about 200 hectares, and Gerard, whose farm did not have the same potential, received 300 hectares.

Fanie set out to develop Cabernet Sauvignon. As there were Cabernet Sauvignon vines that showed good promise as single plants in other red variety vineyards, Fanie selected the best from these and was able to plant the Estate's first Cabernet Sauvignon vineyard in 1962. Allesverloren is situated east of Malmesbury, some fifty kilometres from the sea. "I know that we're a long way from the sea," says Fanie, "and I believe that the theory that good red wines can only be made near the sea is just that, just a theory".

Fanie Malan has switched the emphasis of the farm from port wine production to red table wine. In addition to the red varieties his father had planted for port wines, Fanie has planted Carignan and Pinotage. The farm now has nine red cultivars, all suitable for dry red table wine production. In addition to these, Fanie still has red Muscadel, from which he makes sweet red wine. One third of the 135 hectares of vineyards is planted with white cultivars, and the rest with red.

The one characteristic of all the soil on the Estate is its gravelly, pebbly structure, which may be the reason for the extraordinary depth that vine roots reach on Allesverloren.

The soil and rainfall differ all over the farm. Allesverloren stretches down the mountainside from a high point about 300 metres above sea level to vineyards that are less than 130 metres above sea level. The top part of the farm receives 100 mm per year more rain than the middle section, where the cellar is situated, and 200 mm per year more rain than the lowest part of the property. The soil varies from an alluvial sandy loam, to clay, to various mixtures of sand and clay. This and the variation in rainfall mean that the choice of the right variety for the right vineyard is a painstaking, careful process.

Pruning and training new growth along Allesverloren's trellising is an important winter task.

To replace old vineyards and to increase the production of quality red wines, Fanie has planted more Shiraz, Cabernet Sauvignon, Tinta Barocca, and Souzão vineyards. The first two varieties are scheduled for the production of more table wines, while the Tinta Barocca and Souzão vines will produce more grapes for Allesverloren port. Fanie has planted an extensive vineyard of Sauvignon blanc, but makes no white wines in his cellar. The white grapes are sold to a wine merchant in Stellenbosch.

Until the late '70s, Fanie irrigated his vineyards to supplement the mainly winter rainfall, but found that even judicious addition of water was detrimental to wine quality.

The old vines that Fanie acquired were all bush vines. When he took out the bush vine Shiraz vineyards, he trellised the Shiraz that was replanted in its place. All new vineyards either have been or will be trellised. Tinta Barocca, of which Fanie Malan has more stock than any other red variety, is best grown on a low trellis.

Allesverloren is one of the few Estates to have experimented with heat extraction of colour and flavour from

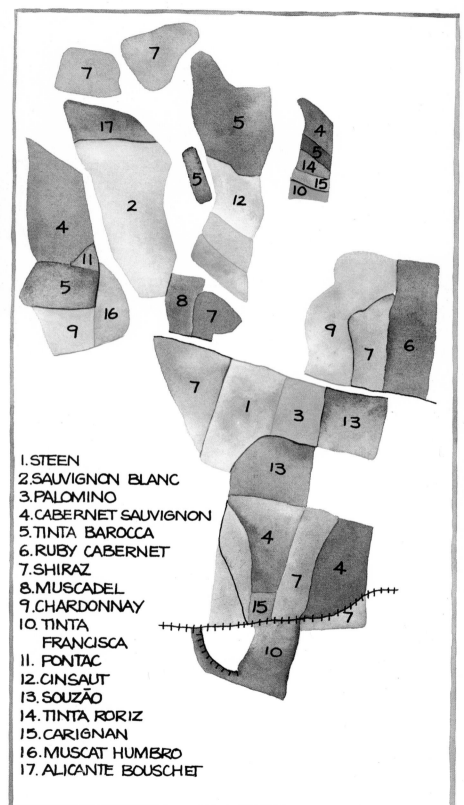

1. STEEN
2. SAUVIGNON BLANC
3. PALOMINO
4. CABERNET SAUVIGNON
5. TINTA BAROCCA
6. RUBY CABERNET
7. SHIRAZ
8. MUSCADEL
9. CHARDONNAY
10. TINTA FRANCISCA
11. PONTAC
12. CINSAUT
13. SOUZÃO
14. TINTA RORIZ
15. CARIGNAN
16. MUSCAT HUMBRO
17. ALICANTE BOUSCHET

Allesverloren

Allesverloren's vineyards begin almost at the foot of the rocks on the slopes of Kasteelberg.

Allesverloren's highest and lowest vineyards vary in elevation by more than 170 metres.

Deep ploughing allows Allesverloren's vines to obtain a good start, with roots able to thrust deep from the first year.

red wine skins. All the other procedures involve leaving the juice with the skins for a matter of days, and use various means of mixing the fermenting liquid must with the skins to extract colour and flavour. Fanie used his first piece of sophisticated heat extraction equipment in 1974 on a very small section of his red wine crop. He found that the wines made using heat to extract colour were lacking in body and tannins and were unable to benefit from a period of maturation in oak.

He has therefore reverted to the exclusive use of the traditional method of fermenting in open concrete tanks with the skins of the grapes in the fermenting must. In the past he held the temperature of the must down to approximately 19° Balling. More recently he has allowed the must to ferment at a slightly higher temperature and the skins have been retained in the mixture until the must is approaching the point of dryness. The wines show a greater depth of flavour and vitality than earlier wines at a similar age. A selected vineyard of Cabernet Sauvignon is used to produce a full-bodied, richly-flavoured pure cultivar wine. "But I'm not particularly keen on these pure cultivar wines," says Fanie. "I prefer a blended wine and we'll be putting a great deal of effort into the Swartland Rood. It'll continue to be a Shiraz-based blend, but I'll make sure that our best wines from all cultivars will be ready to go into the mix, if they're needed."

Fanie's style has been set by the unique location and natural benefits of his farm. Grapes ripen easily, reaching high concentrations of sugar when left on the vine. The vineyards produce grapes that make full-bodied wines. These reserves of sugar have been responsible for the Estate's long history of quality port production. Like his father, Fanie is a renowned port producer. His ports are accepted as some of the best of their type made in South Africa. He is trying to move his port style closer to that of the Portuguese original. "I want to get more colour in my ports, with less sacrifice of acidity," he says. "I want to pick the grapes a little earlier, when they have less raisin character. I've planted much more Souzão, as the grapes have higher acidity and more pigment in the skins."

The Allesverloren red wines are fermented in the Estate cellar and then transported to Stellenbosch, where they are matured, bottled and nationally distributed by a wholesaler.

Wine is not for sale on the Estate.

TULBAGH

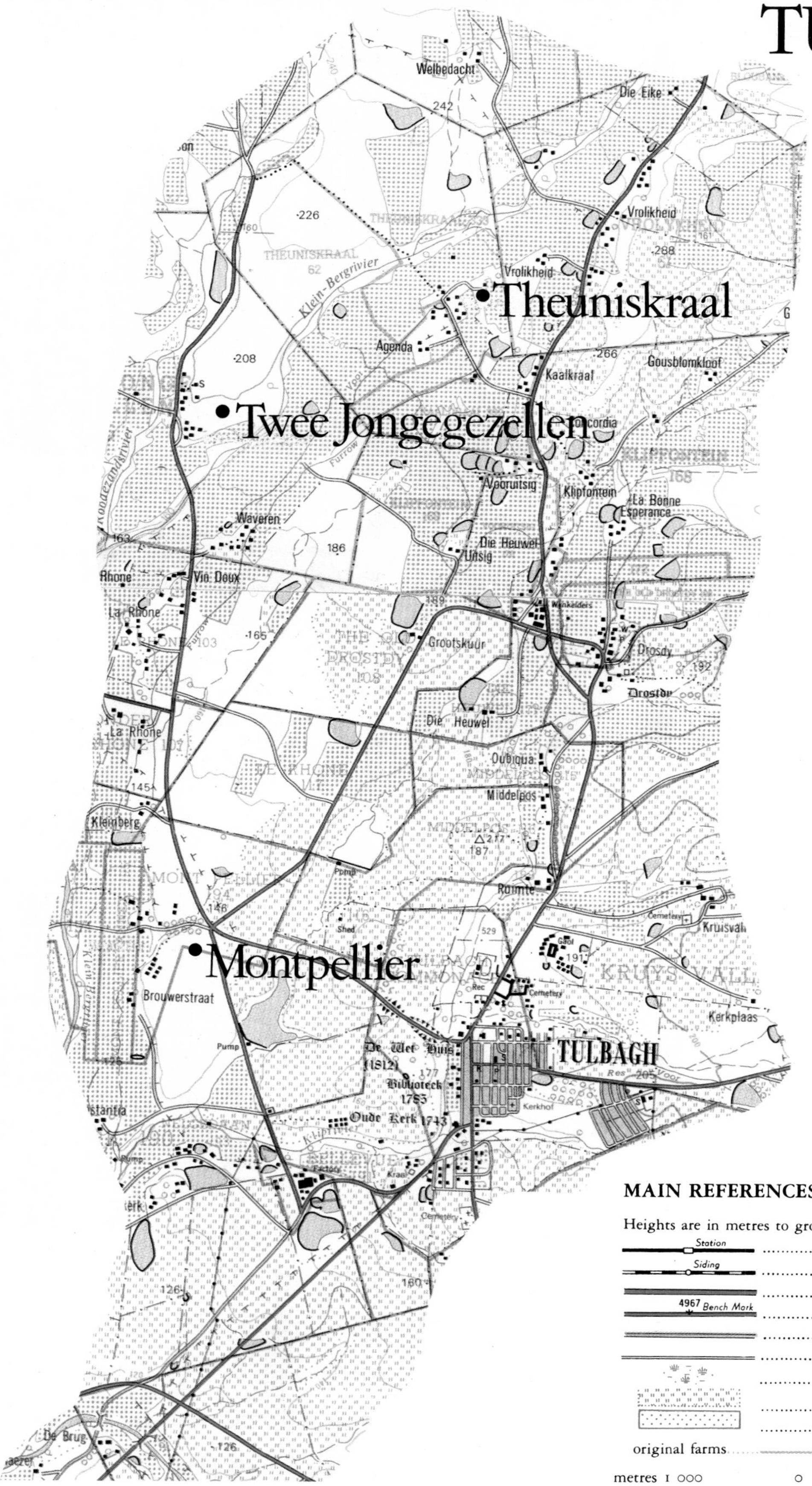

- Theuniskraal
- Twee Jongegezellen
- Montpellier

MAIN REFERENCES

Heights are in metres to ground level

.... multiple track railways
.... single track railways
.... national roads
.... main roads
.... secondary roads
.... other roads
.... marshes, swamps and vleis
.... cultivated lands
.... orchards and vineyards
original farms

metres 1 000   0   1 000   2 000

Map by permission Government Printer, Pretoria

Montpellier

De Wet Theron, who rebuilt his historic home with such attention to detail that it has become one of the most gracious homes of the Cape and has been declared a national monument, also rebuilt his wine farm and cellar so effectively that the farm is one of South Africa's top white wine Estates.

The Theron house on Montpellier dates from the very early years of the eighteenth century. It was almost totally destroyed by the 1969 earthquake that shattered the small wine-farming area around Tulbagh. Much of the village has been restored, but many buildings were damaged beyond repair. After the earthquake, the Therons on Montpellier were faced with a choice: either build a new house or rebuild the old one. They chose the latter course and methodically proceeded to take apart the parts of the house still standing. This enabled them eventually to form a detailed plan of how the house had developed in size and style over the centuries.

The buildings had obviously been the original home of the first owner of Montpellier, Jean Joubert. Joubert was granted 50 hectares in the Tulbagh valley by De Chavonnes, then governor of the Cape Colony, in 1714. He named it Montpellier after his home town in France. From the sod wall and mud wall construction of the first rooms of the old homestead, discovered when the remains of the old house were dismantled, it may be assumed that Joubert had lived on the land for some years before he obtained the freehold grant.

The first part of the house was a three-roomed building with a kitchen, living-room and bedroom. The house was first elongated and later a wing was added to the centre. This wing became the new kitchen. Later the wing had added to it another room, which in turn became the kitchen, the old kitchen becoming a dining-room. At an even later stage a second wing was added to the first and the final addition was a room that became and has remained Montpellier's fourth kitchen.

Severely damaged in 1969 by a major earthquake, the Montpellier house was rebuilt by the Theron family.

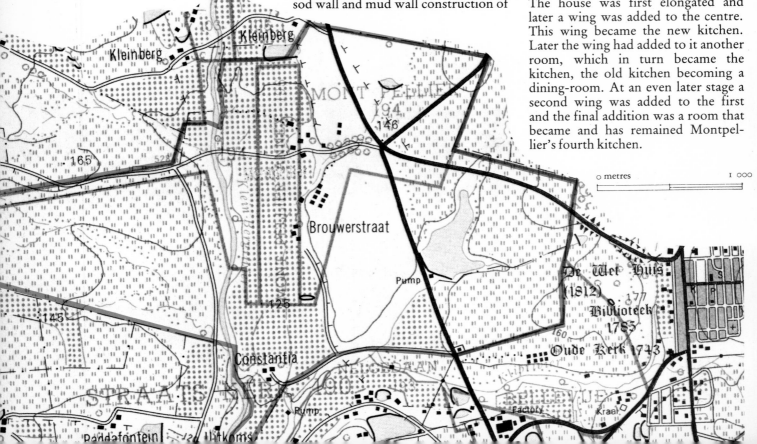

Maurits Pasques de Chavonnes, the Governor of the Cape who granted Montpellier to Jean Joubert in 1714.

De Wet Theron's rebuilding of his wine farm has been a much longer process, developing the vineyards and the cellar. When he took over the management of Montpellier in 1945, he found only Sémillon (then known throughout South Africa as Green Grape), Riesling, white French and Cinsaut. His training, first at the University of Stellenbosch, where he did three years post-graduate study in viticulture, and later at the KWV, where he worked as Dr. Perold's assistant and, after Dr. Perold's death, as wine and brandy expert, encouraged higher ideals for the farm.

On his return to the farm he immediately removed the Cinsaut vines, believing Montpellier to be totally unsuited to red wine making. Most of the other vines were old and were replaced by Clairette blanche, Colombard, Riesling and Steen vines.

Much greater changes came later. In 1954 a leading South African wine merchant brought Gerhard Kreft, a wine expert, to the country. Kreft was to experiment with the production of high-quality South African white wines in particular, and to install the Geisz system of pressure fermentation, which had been very successful in Luxembourg. Working in collaboration with the wine company, Montpellier was to become the site of much of Kreft's experimentation and development. But the pressure fermentation project was not a success on Montpellier. The wines were flat in taste, and did not improve with maturity.

Then De Wet himself began to experiment with the problems of heat and oxidation in the production of white wine. In 1958, soon after the Kreft project had finished, a new procedure of running cold water over the tanks to cool the must as it fermented was introduced.

The first moves towards bottling Montpellier's Riesling started with a renewed agreement with the same Stellenbosch-based wine merchant. Another leading cellar master had been brought to South Africa to improve the quality of white wines. Karl Werner chose Montpellier as the base for his experiments towards the reduction of oxidation. He was responsible for the effective use of CO_2 in the cellar, the rapid cooling of grape juice and the general control of temperature throughout the cellar. On Werner's advice, De Wet rebuilt his cellar in 1965. New machinery and equipment were installed and every method of reducing oxidation was applied to cellar procedure.

The farm adopted the practice of picking in 20-kilogram plastic lug boxes to prevent premature crushing of grapes handled in bulk. Handling procedures were changed to reduce to the minimum the time taken, from the cutting of the bunches in the vineyards to the cooling of the crushed grape juice in the cellar. And at all stages in the cellar the juice and must were protected as much as possible from the oxygenated atmosphere by CO_2.

During Werner's time on Montpellier there was so much innovation and invention that the cellar in 1967 bore little resemblance to the cellar of 1963. From 1967 a selected culture of yeast has been imported for each season's fermentation. Many Estates used sulphur during harvesting and crushing, but since Karl Werner's time Montpellier white wines meet sulphur only after they have been fer-

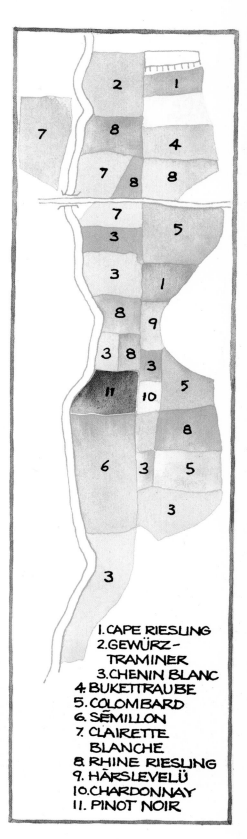

1. CAPE RIESLING
2. GEWÜRZTRAMINER
3. CHENIN BLANC
4. BUKETTRAUBE
5. COLOMBARD
6. SÉMILLON
7. CLAIRETTE BLANCHE
8. RHINE RIESLING
9. HÁRSLEVELÜ
10. CHARDONNAY
11. PINOT NOIR

Montpellier

mented dry, when SO_2 is used to assist in stabilisation before bottling.

In 1967 Karl Werner persuaded De Wet to bottle his own wine. On 2 December 1967 Cape Riesling from the '67 crop was bottled. This was the birth of Montpellier Estate wines. One thousand bottles were bottled by hand, corked by hand and laid down for maturation.

In 1968 Werner returned to Germany. De Wet bottled part of the '68 Riesling crop, though none of this wine was ever sold. It was kept for checking and used for discussion with South African and visiting wine experts. The first four years of bottled Riesling were kept purely as experimental products until 1971, when it was agreed that bottle maturation of Montpellier white wines was both worthwhile and beneficial.

The '71 crop of Riesling was bottled and kept for eighteen months before sale. That wine, in 1973, became the first South African white wine to

De Wet Theron, once wine and brandy expert at the KWV, and his grandson. De Wet runs Montpellier with his sons Jan and Hendrik.

be classified 'Superior'.

De Wet's second son, Jan, was in the enviable position of being able to serve virtually a three-year apprenticeship in cellar procedure under Karl Werner, without having to move from his home conditions. He could not have wished for a better teacher.

The most recent addition to Montpellier's cellar has been the introduction of a bottle maturation area that is kept at a constant temperature.

The rebuilding of Montpellier wines was not limited to the cellar. We have mentioned the work that has been done with the Estate Riesling. But during this time De Wet was not happy with the quality of many of the varieties replanted in the years after 1945.

As Riesling had proved that better varieties gave a recognisably better product, De Wet looked to find other cultivars to develop the Estate's production. As the 600-hectare farm has 160 hectares of vineyards, he sought more varieties to provide a range of high-quality white table wines.

The first new cultivar given a trial in the Montpellier vineyards was known at that time as Traminer. De Wet obtained grafted vines of these vines from Elsenburg in the middle 1960s and planted blocks in 1967 and 1969. This variety produces noticeably pink grapes and is known in Alsace and Germany, where it is traditionally grown, as Roter Traminer. The wine it produced was light, more delicate and fruity in flavour than the Cape Riesling, and Montpellier's first

Montpellier's alluvial soil vineyards are on both sides of the Klein Berg River, on the western edge of the Tulbagh Valley.

Montpellier's early harvesting seems suited to the delicate style of Cape Riesling, producing consistent wines.

Weisser Riesling wines made at Montpellier benefit from the Estate's practices of bottle maturation before release.

The rich riverland soils assist the vines to grow vigorously, providing a leafy cover for the ripening bunches.

vintage was bottled in 1971 and certified 'Superior' in 1972. Around this time new clones of this variety were being tested by the OVRI. These were known in Europe as Gewürztraminer, because of the more pronounced spicy flavour in the wine. De Wet was persuaded to label his first and subsequent bottlings with this name.

Jan Theron, one of De Wet's two sons, is the Estate cellar master. The other, Hendrik, is in charge of the farm's vineyards. They have found Traminer to be an extremely variable cultivar. The grapes are not highly endowed with acidity, and the centre of the Tulbagh Valley often has high temperatures in early February when Traminer is ripe, causing the acidity to tumble. Jan has found that only a few of his vintages of Gewürztraminer have kept well in the bottle. Hendrik has lately planted a vineyard of a more recently released clone of Gewürztraminer. "Our customers like the simple style of our Gewürztraminer. So I've planted the new material on lighter soils, aiming to produce a gently flavoured wine that will blend well with the product of our old Traminer vines.

"The next variety that we tried out here was another that varies greatly in the quality of the grapes that it produces each season," says Jan. "We were among the first to plant and make Estate wine from Weisser Riesling. For several seasons we marketed it as Tuinwingerd to avoid confusion with our traditional Riesling. I'm experimenting with wood maturation of Weisser Riesling before bottling. Without wood maturation, the wine develops a 'paraffin' character after about eighteen months that stays for at least another year in the bottle. I think the time in the wood removes this tendency.

"We've since planted Pinot gris, which will probably go into our Late Harvest. I'm very impressed with the wine so far. The other major new variety we've planted is Pinot noir, which will be used exclusively for bottle-fermented sparkling wine.

"I've experimented with giving the juice more skin contact before fermentation, but I've found that the longer the skin contact, the shorter the bottle life. We have too high temperatures for prolonged skin contact."

The two stalwarts of the Montpellier cellar are Cape Riesling and Chenin blanc. These varieties produce delicate wines of a consistent standard in the low-alcohol Montpellier style each year. The vineyards of Colombard and Clairette blanche, once the chief ingredients in Montpellier's lowest-priced blend, have been reduced to allow increased plantings of the more highly prized varieties.

Wine is for sale on the Estate.

Montpellier

Theuniskraal

Mention the name Theuniskraal to a South African wine lover and he is almost certain to say Riesling. For Theuniskraal is one of the most famous Riesling Estates in the country. The farm is situated on flat ground in the north-west corner of the Tulbagh Valley and was once part of the historic La Rhone property. The south-west boundary of the farm is adjacent to Twee Jongegezellen, another famous Tulbagh white wine Estate. Though the farms share a common boundary, the conditions, such as the micro-climate and the environment, under which vines grow on the two farms are very different.

Theuniskraal's soil is alluvial and rich, brought down by rivers from the mountain range just a few kilometres to the north. The farm is flat and the soil easy to cultivate. Because the soil is deep and well drained, irrigation has to be used quite heavily. Probably owing to the fertility of the soil this quantity of irrigation does not seem to affect the quality of Theuniskraal wine.

Until the mid 1970s the Theuniskraal Estate specialised almost exclusively in the production of Riesling wine, but since the farm has been run by the two brothers Kobus and Rennie Jordaan, a partnership that began in 1964, careful selection of a number of other cultivars has resulted in an extended range of wines bearing the Theuniskraal label. Cultivar wines under the names of Steen, Sémillon and Gewürztraminer have been added to the range.

Theuniskraal is not a traditional family farm in the way of many South African wine Estates, which have often been owned by the same family for periods of as long as 200 years. Theuniskraal is owned by the Jordaan family, comparative newcomers to the Tulbagh area. The history of the Jordaans is to be found across the mountains in the Hex River Valley, where they produced table grapes. Any wine making they did before they came to Tulbagh in the 1930s was of a domestic nature.

A. W. Jordaan's mother bought Theuniskraal in 1930, with the intention of producing high-quality white table wines. She intended to give her a son a farm capable of becoming a wine Estate of note, for the Tulbagh Valley was already noted for the quality of its white wine production. This was in the years before cold fermentation and other cellar techniques revolutionised South African white wine making. Mr Jordaan found his new farm to be stocked with Green Grape (Sémillon), Steen and a small area of Cape Riesling. He developed the Riesling vineyards to the extent that when this wine first went on sale to the public in 1948 it was an immediate success. Riesling was Theuniskraal's only Estate wine for many years. The Sémillon and Steen production was made into wine for sale to wine merchants for blending with other wines and distillation.

This practice continued unchanged until 1964, when Mr. Jordaan's sons, Kobus and Rennie, formed a partnership to manage the farm. Kobus became cellar master, and Rennie took charge of the vineyard. They started a process of visual selection of the better vines of each variety. From this beginning they were able to build up an area of both Sémillon and Steen vines that produced light, delicate table wines of sufficient quality to bear the Theuniskraal label.

The first new variety to be planted by Rennie Jordaan in Theuniskraal soils was Gewürztraminer, a quality variety grown in Alsace and Germany. Traminer had previously been tried in the Stellenbosch area, but Rennie's first commercial crop of Gewürztraminer started a wave of plantings of this variety. On Theuniskraal Gewürztraminer is allowed to ripen fully, producing the heavily scented, spicy-flavoured wine that has given the variety its name. Several years of bottle maturation normally give more concentrated flavour.

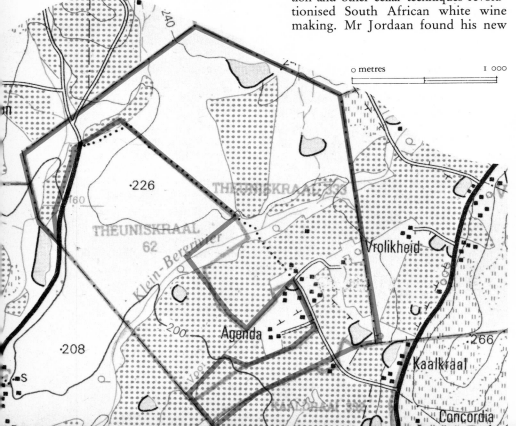

After the last drop of the juice has been pressed from the grape skins, the residue is returned to the vineyards.

Rennie has lately planted vineyards of Pinot gris, Weisser Riesling, Muscat Ottonel and Sauvignon blanc. Kobus Jordaan, the wine maker, has found Weisser Riesling to be an excellent blending partner for Cape Riesling, founder of the Estate's great name for white wines. "We still pick our Riesling early, for extra lightness and fragrance," says Kobus, "but we are giving our white juices longer periods of contact with the skins to add complexity to the flavour. The other new varieties, in particular Sauvignon blanc, will be examined with interest, for only if we produce something really outstanding will we add more labels to the range. We're not growing enough Riesling or Gewürztraminer yet to satisfy the demand."

Theuniskraal lies on the flood plain of the Klein Berg River, and the vineyards have a thick covering of water-rounded stones that have traditionally been a hindrance to vineyard cultivation. The Jordaan brothers now crush these stones, producing fragments that make a reflective cover for the vineyard surface, which lowers the temperature of the soil around the vine roots.

Theuniskraal's deep, fertile soils require the use of extensive sprinkler irrigation.

Kobus Jordaan, the cellar master, who with his brother Rennie, has operated Theuniskraal since 1964.

Theuniskraal lies in a valley between the Obickwa Mountains (shown) and the Witzenberg range.

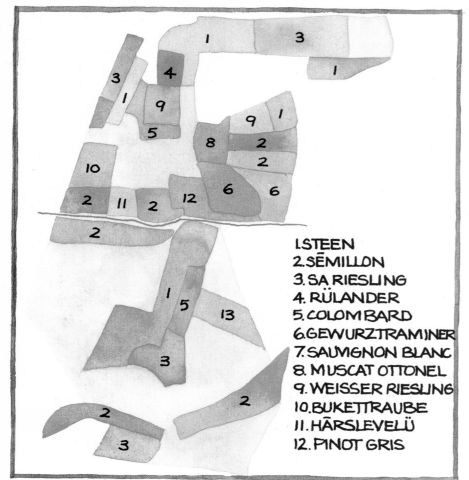

1. STEEN
2. SÉMILLON
3. SA RIESLING
4. RÜLANDER
5. COLOMBARD
6. GEWURZTRAMINER
7. SAUVIGNON BLANC
8. MUSCAT OTTONEL
9. WEISSER RIESLING
10. BUKETTRAUBE
11. HÄRSLEVELÜ
12. PINOT GRIS

The Estate was one of the first in South Africa to use a mechanical harvester, which operates in those vineyards that are suitably trellised. The harvester starts collecting grapes and juice at 5 a.m. and finishes for the day at 10 a.m. At present about 15 per cent of the production is picked in this way and no significant difference in quality between the different harvesting methods has been noticed.

The Theuniskraal Estate has an agreement with the Bergkelder, which undertakes to bottle the Estate's produce.

Wine is not for sale on the Estate.

Theuniskraal

Twee Jongegezellen

Though Twee Jongegezellen has switched to night-picking, 10 per cent of the crop is still harvested by day.

The day team cuts off the imperfect bunches, allowing the night-picking group to harvest without pause.

"I'm looking forward eagerly to the day when South African white wines are really in the top rank," says Nicky Krone. "I don't think we're far away. We're finally on the right track. We lost our way a bit when we discovered that cold fermentation could reproduce some of the conditions under which the good wines of Europe are produced. We forgot that wines are not made in cellars. The cellar is like the maternity home. The wine is made in the vineyard and we simply deliver it in the cellar.

"What we've been missing all along has been the correct genetic material in our vineyards. In Europe, they're continually selecting and upgrading the classic varieties, and constantly improving their vineyards. We've been chasing fondly after varieties that'll give more production than these classic vines, and hoping we can turn them into wonderful wines in the cellar.

"On Twee Jongegezellen, we've turned our attention back to the vineyards. I have 34 different varieties here and I'm trying to find out something from each one of them. But I've come to the conclusion that there are only 5 classic white cultivars, and if I can get those established here, producing their best according to our natural limitations and advantages, I'll have done a good job. We'll be concentrating on Chardonnay, Sauvignon blanc, Gewürztraminer, Rhine Riesling and Sémillon.

"I've chosen to put Sémillon (*not* Green Grape, which is a mutation of the classic variety) in my top group, just as my father did some years ago, because, unlike South Africans, the French rate this variety highly. It is the only white variety grown by Château Haut Brion, makers of possibly the best white wines in the whole Bordeaux area. Because their cellar produces one of the most famous clarets in the world, their white wine is rather overshadowed, but it's still a marvellous wine."

Before 1700 Twee Jongegezellen was a cattle-post rented by La Rhone to two bachelors, who planted the first gardens and ran the first wheat mill. There is no record of the names of these *twee jongegezellen* (i.e. 'two young companions'). In 1745 a daughter of Theron of La Rhone married a Du Preez, and was given Twee Jongegezellen as a wedding gift. The inventory value in 1745 was 30 rixdollars. In 1710 the first house was built in the cattle kraal. The walls of this were destroyed only in the earthquake of 1969, as were the remains of the mill. In 1719 there was begun, in the traditional Tulbagh T-shape, a bigger house, which forms the heart of the present homestead. The gable was probably added later, the date on it referring to the actual origin of the *opstal*. The farm has never passed out of the family, the present owners being direct descendants of the original Therons.

In the early 1900s N. C. Krone, whose family owned a wholesale liquor business in Worcester, married a daughter of C. J. Theron, owner of Twee Jongegezellen. Soon afterwards he bought the farm from his father-in-law.

At this time Twee Jongegezellen was well stocked with healthy vines, in contrast to the orchards and stricken vineyards of most of the rest

In 1982, Nicky Krone began to harvest 90 per cent of the Estate's grapes at night to obtain cooler, fresher juice.

Twee Jongegezellen has 34 different varieties in the Estate vineyards, mostly white. All must be crushed and fermented separately.

of the Cape. The Theron family vineyards had been ravaged by Phylloxera during the 1890s, but new vineyards were soon planted using American resistant rootstock grafted on to French and Spanish grape-bearing vines. In 1951 N. C. Krone's son, also named Nicolas Christian, took over the property and lifted the Estate's produce to the forefront of the Cape wine industry. Today Twee Jongegezellen is managed by a third N. C. Krone, known as Nicky, and innovative practices continue.

Twee Jongegezellen's 240 hectares of vineyards are situated on the western edge of the Tulbagh basin. The farm starts along the alluvial soils of the northern bank of the Klein Berg River and runs up the slopes of the mountain range to the north west. Most of the Estate's best vineyards lie on the recently developed slopes at the foot of the Obiqua mountain. The slopes are in the rain shadow of the mountain and receive higher rainfall than other parts of the Tulbagh area. The soils here are thoroughly mixed with small and large stones and are much more difficult to cultivate than the alluvial soils in the valley below. To plant vineyards on these slopes, the ground has to be terraced. Adding to the difficulty is a layer of ferrocrete conglomerate stone a metre or so under the terraces, which has to be deep-ripped to break up the stone and allow the vines' roots to penetrate.

The sloping ground faces east and compensates the vineyards for the relatively high temperatures recorded in the Tulbagh Valley. Vines on the slopes receive the early morning sun while the dew is still on the leaves. When the sun sets in the evening behind the Ou Kloof mountain, these vineyards fall into shadow before the vineyards on level ground.

Tulbagh is one of the most northerly of the recognised quality wine areas in South Africa, and summer temperatures are normally high in this sheltered valley. To minimise extreme effects of the summer sun, Nicky has positioned all his recently planted vineyards with rows running from east to west. The vines' summer canes are trained to keep the bunches out of direct sunlight until they are harvested. Sunburnt grapes have always presented problems in Cape cellars, as a certain amount of oxidation is inevitable once the berries are damaged.

Though Twee Jongegezellen is in the coolest part of the valley, Nicky Krone believes that the quality of his wines improves when the grapes are picked at the lowest temperatures. He experimented with night-picking small batches of grapes between 1977 and 1981 and compared the analyses of juice obtained from night-picked grapes against juice from comparable grapes harvested during daylight hours. Under hot conditions the vine loses a great deal of moisture through transpiration. It is believed that the vine strains to replace the decreasing moisture and uses extra quantities of nutrition that might have been used instead to build flavour compounds and sugars in the grape.

"During the night the vine doesn't sleep," says Nicky. "It tends to revive and the berries become crisp and fresh, intensifying the flavour substances, particularly near the skin. During the day it loses water through transpiration to the atmosphere, as well as some of the more volatile constituents of the juice, which it draws from the berries when it's under stress. These would normally play a major part in the bouquet and flavour of the wine that we'd make from the juice. Our tests have shown that grapes picked in the middle of the

The Krone family's historic home was severely damaged by the 1969 earthquake. It was restored and declared a national monument.

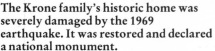

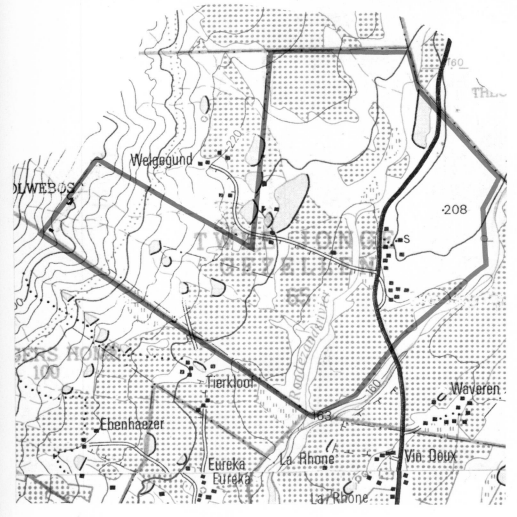

night contain more of everything than grapes picked from the same vine in the middle of the day.

"High temperatures dissolve polyphenols in the juice. The best wines of northern Europe don't have these problems, and because local temperatures at night time are more like those of northern Europe than the temperatures of our days we've gone over to night-picking about 90 per cent of the crop. Some of our vineyards planted on the stony soil of the mountainside have patches of uneven growth. A few of the vines have an insufficient cover of leaves. To maintain a constant level of quality, I send a team out in the day time to pick out any bunch that may be inferior and these are crushed and their juice is fermented separately. When the night-picking team go into that vineyard, they're picking consistently high quality grapes. To emphasise the difference between day and night, the warmest temperature of juice produced by grapes picked during the day was 37 °C. It was a real scorcher that day. At night the temperature dropped in the normal manner, but the highest juice temperature recorded after 11 p.m. was 18 °C. Grapes picked in the day time give brown juice. Grapes picked at night give green juice. A certain amount of oxidation is inevitable at higher temperatures and I'll do almost anything to prevent this, because oxidation breaks down the amino-acids that are the flavour-building substances in the juice. Clean, fresh, greenish juice makes fresh, flavourful wine."

The night-harvesting teams are equipped with waterproof jackets and trousers to combat the dew, and wear lightweight battery packs to power the miners' lamps on their helmets. They work at a faster rate than their counterparts during the day. The temperature is lower and the lights clearly illuminate the bunches. The first teams go into the vineyards at about 8 p.m. each night and the last loads come into the cellar around 7 a.m. The cellar is operational for 24 hours a day during this period, but night-picking removes some of the crisis management involved in the op-

N. C. Krone (right), one of South Africa's most famous winemakers, and his son Nicky, the present vintner in the TJ cellar.

Twee Jongegezellen was first farmed shortly before 1720 and has never been sold, passing on from father to son or daughter ever since.

eration of a cold fermentation cellar. High-temperature juices have to be crash-cooled for preservation and to settle out solids before fermentation. With lower-temperature juices coming from the crusher, the cooling facilities can be used more efficiently, to maintain other quality-related procedures, such as regular rates of fermentation and efficient stabilisation. The lower juice temperatures allow Nicky to increase the period in which the skins of his white grapes soak in the juice before fermentation, providing increased flavour.

About 20 per cent of the juice used for the fermentation of the top wines comes from a final squeeze of the skins in a basket press. This is added to the first free-run juice. The juice is chilled, the solid matter settled and the clear liquid drained off.

Nicky uses multiple cultures of yeast to ensure vigorous, sustained fermentation, even at relatively low temperatures. He believes that the yeast contributes to the flavour of the wine and that multiple yeast strains add to the complexity of the wine's flavour.

"Europe's top wines are made with the yeasts found in their vineyards and cellars. Each load of grapes carries many different strains of local yeast and all of these operate efficiently in the low temperatures of their cellars. When we introduced cold fermentation, we found that our vineyard yeasts could not operate efficiently at such temperatures, and we had to import European yeasts that would reduce sugar at 10 °C. But we've tended to use one yeast at a time, and we believe that monocultures of yeast have contributed to the neutral flavours of our wines.

"We now obtain a number of yeasts well before the harvest and start them fermenting slowly. We choose about half-a-dozen of the most active and prepare our inoculations with this group. We keep them separate right up to the moment of introduction into the cold settled juice. When those varied strains of yeast, already multiplying at 8 °C, meet the sugary juice, the fermentation begins at once and maintains such vigour that I sometimes have to take the temperature down from 12-14 °C, my normal range, to 8 °C, to keep it running at an even, controlled rate."

Nicky prefers to give the wines the finings that they may require during the period of fermentation. "We try to handle and move our wines as little as possible. When fermentation is completed, we rack the wine off the lees and then hold it at a cold temperature until it's ready for bottling."

The Krone family have enjoyed a measure of success in the young-wine shows in the Cape down through the years. Each wine region has a competition for wines from the previous harvest. Success in the regional shows allows the wines entry to the national

Twee Jongegezellen

The Twee Jongegezellen Estate is tucked away in the north-west corner of the Tulbagh valley, on the lower slopes of the Obickwa mountain. Recent vineyards have been planted on higher slopes.

show, normally held in October, seven months after the last grapes have been picked.

The Estate's most famous show result caused the wine-producing industry in the Cape to reassess some basic quality principles, which has resulted in the development of the extremely high degrees of mechanisation and cellar technology that may be seen in Cape cellars today. In 1959, after an absence from the show circuit of several years, N. C. Krone exhibited fifteen wines in thirteen classes and won thirteen different first prizes. The Twee Jongegezellen cellar, and one or two others, had been using refrigeration to cool juice and fermentation for several years. After this public demonstration of quality, the technique spread rapidly throughout the Cape.

In recent years Nicky has often been successful with young wines in competitions and, in addition, the Estate has become a major producer of blended and pure cultivar white wines on the national market. No fewer than ten white wines carry the Estate label. One of Twee Jongegezellen's major successes has been the introduction of Schanderl, a fruity, off-dry wine, a blend based on Rhine Riesling and Muscat de Frontignan. TJ 39, a blend of 17 varieties combining both the delicacy of the finer German wines and the refreshing dryness of the French, has become the Estate's top seller and one of the country's summer favourites.

Wine is not for sale on Twee Jongegezellen, as the Estate's produce is marketed by a national wholesaler.

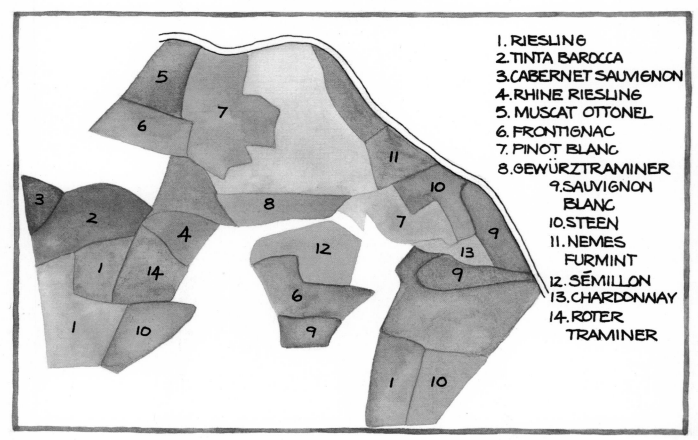

1. RIESLING
2. TINTA BAROCCA
3. CABERNET SAUVIGNON
4. RHINE RIESLING
5. MUSCAT OTTONEL
6. FRONTIGNAC
7. PINOT BLANC
8. GEWÜRZTRAMINER
9. SAUVIGNON BLANC
10. STEEN
11. NEMES FURMINT
12. SÉMILLON
13. CHARDONNAY
14. ROTER TRAMINER

Hamilton Russell Vineyards

The Hamilton Russell vineyards are situated in the Hemel-en-Aarde Valley, just a few kilometres north-west of Hermanus and the ocean. This area, known as Walker Bay, and generally accepted to be the most southerly area on the African continent likely to sustain vines and ripen grapes, contains three main valleys where agriculture is possible.

Although Walker Bay is not much further south than Constantia or Stellenbosch — half a degree of latitude — the proximity of the cold waters of the southern ocean gives the area a markedly different climate and provides vines with a more difficult home in which to live. The prevailing summer wind in the Cape is the south-easter and this blows directly off the water and over the Hemel-en-Aarde vineyards. These extremes naturally make farming considerably more arduous than it is in more temperate locations and explain why vineyards in the Walker Bay area are not more plentiful today.

Before the advent of dams on rivers with seasonal flows and later forms of mechanical irrigation, vines were only grown in areas where there was sufficient rainfall. Stellenbosch, Paarl and Constantia were the most important of these, and Walker Bay and the Elgin area were next in priority. Walker Bay never reached a higher rank, because a high level of quality was never a major factor in the earning of income from wine production in the first 300 years of the Cape wine industry.

When the fertile soils of the Breede River Valley were opened for irrigated vineyards by the construction of a major storage dam on the river in the 1920s, any remaining reasons to struggle with the elements by the seaside disappeared and the vineyards were gradually replaced by orchards, dairy farms and pastures for grazing. The reasons for the existence of the original vineyards and their subsequent disappearance were chiefly economic. Today, the tide has turned once more and changed economics in the wine industry have made the expansion of vineyards in these valleys most likely.

Today there is a developing fascination amongst wine lovers with the extreme south. This may be explained by the story of Tim Hamilton Russell and his quest for the ultimate South African wine. Like most wine lovers, Tim believes that the finest wines in the world are produced by the French and the Germans, and that the cool temperatures in their vineyards, promoting a long ripening period, are chiefly responsible. He realised that Stellenbosch was one of the most popular areas in South Africa for vine growing because of relative warmth and lack of extremes in climate, and began to search for something better. The apple-growing areas around Elgin offered the required temperature but they were insufficiently far south. Walker Bay was almost the only answer. You can travel another half a degree further south, as far as Cape Agulhas, but the sand dunes which constitute the soil of that countryside are unlikely to support a crop of grapes.

On Christmas Day 1974 Tim bought two farms in the Hemel-en-Aarde Valley, one west-facing, just above the village of Hermanus, and the other south-facing, on the elevated slopes of the Tower of Babel Mountains. According to measurements taken to date, and applied to the Winkler heat summation scale of

The vineyard in the foreground contains Sauvignon blanc, and the vines behind the home are Pinot noir and Chardonnay.

The valley vineyards to the left of the cellar face west. The mountain vineyards on the slopes in the distance face east.

vine-growing climate zones (summarised in respect of South African vine-growing areas on page 28), both Tim's farms qualify for classification as Region II.

The mountain vineyards (south-facing) are planted in soils of decomposed granite that are fairly well drained and sometimes need supplementary water, particularly for young vines. The valley vineyards (west-facing) are planted in decomposed sandstone and arenaceous shale soils, with a base of clay not far beneath the surface.

When Tim Hamilton Russell bought the farms, his first idea was to grow Cabernet Sauvignon, Merlot, Cabernet franc and Malbec in the mountain vineyards and Pinot noir and Chardonnay in the valley vineyards. The mountain farm has been found too cold for the Cabernet Sauvignon vines to ripen sufficiently, but it provides good growing conditions for Chardonnay, which is replacing the Cabernet Sauvignon. In time, Hamilton Russell Vineyards will produce only two major wines, a pure Pinot Noir and a pure Chardonnay. A small experimental patch of Sauvignon blanc vines has proved so well suited to the conditions that Sauvignon Blanc wines may also become part of the programme.

When a vineyard is prepared for planting, the soil is deep-ploughed to break the surface of the clay formation and create a mixture of clay and soil where the vines' roots normally have to stop short. The acid nature of the soil is modified by the addition of lime, which is thoroughly mixed with the top soil.

As the project has the single objective of producing superior quality wines, over-vigorous development of the young vines is restricted to encourage greater root and trunk development. Their first few crops are thinned out to reduce strain on the plant. Mature vines are also pruned to prevent the crop exceeding seven tonnes per hectare.

The decision to pick the grapes is determined by the pH content of the grapes as they ripen. Peter Finlayson, whose previous cellar experience includes Montagne and Boschendal and who is the Hamilton Russell cellar master, tries to have all the grapes harvested before they reach a pH reading of 3,3. In general, this means that they reach the cellar with a Balling measure of 21-22° and with considerable natural acidity.

The juice of the Pinot noir grapes is fermented in a closed tank at a relatively low temperature with all the skins and a small quantity of stalks until the wine is dry. After racking to lightly clean the wine, it is placed in small oak casks for 12-18 months, and then bottled. The first Chardonnay wines were treated to 24 hours' skin contact, followed by fermentation, partly in wood, at 20 °C. The Sauvignon blanc grapes are pressed hard in a basket press to extract the utmost flavour and then chilled and fermented at about 12 °C without having had any appreciable skin contact.

The wines are distinctive and original and seem to be very much worth the trouble. It is likely that Tim's idealism will produce a very successful enterprise.

Wine is for sale at the Cypress Wine Tavern in Hermanus, and at certain periods during the year is available from the more discriminating retailers.

Peter Finlayson shows the successful union between the Chardonnay cuttings and the rootstock.

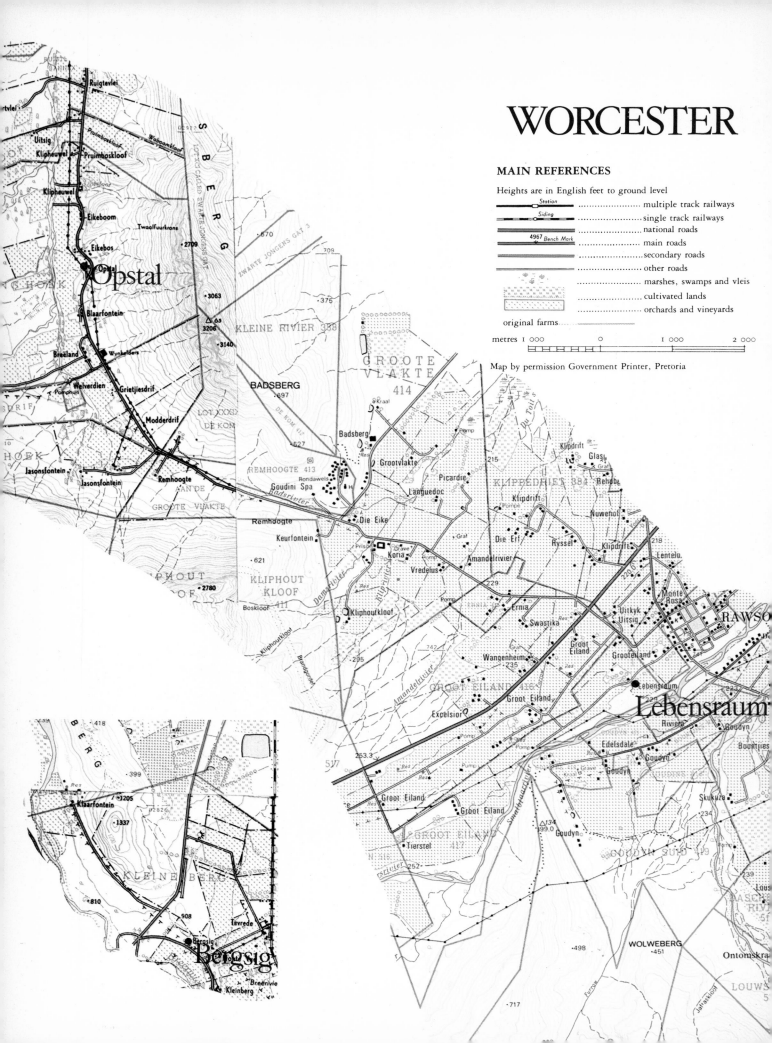

Bergsig

Situated at the foot of the Bainskloof Pass, in the narrow channel where the Breede River flows between the Mostertshoek Mountains and the Elandskloof Mountains, Bergsig is exposed to the southerly summer winds that damage vines, lower temperatures and bring moisture to certain parts of the Cape during the warmest and driest part of the year. Most of the regions that benefit from the summer breezes are closer to the coast, but Bergsig benefits from the funnelling effect of the mountains. The Estate benefits also from 700 mm of yearly rainfall and moderate temperature by ripening the crop of 400 hectares of vines over a period of two months and producing a large volume of wines of consistent quality. Though the rainfall is above average for the inland areas of the Cape, all the Bergsig vineyards are irrigated.

Bergsig is a large property, covering more than 1 000 hectares, approximately half of which are mountainous and impossible to cultivate. The level and gently sloping parts of the farm are divided into three sections. Along the Breede River, the soils are sandy in type, deep in extent and their relatively high fertility produces large crops of grapes. On higher ground, beside granite outcrops of rock, Hutton-type soils with relatively high clay content provide smaller yields of higher quality grapes. The third section of the Estate benefits from the higher pH of Karoo soils and is able to provide higher production with a reasonable standard of quality. The Worcester fault, a geological break in the earth's crust, running west from Worcester to the Atlantic coastline, cuts through the Estate and separates the Karoo soils from the sandstone-derived soils. The consequent wide disparity between soil types and growing conditions allows Prop Lategan, Bergsig's owner, to stretch out his harvest of Chenin blanc, at sugar concentrations of between 19 and 25 ° Balling over a period of three weeks.

Unique in situation, with relatively cool growing and ripening conditions in the warm Worcester Valley, Bergsig produces grapes with a gently-flavoured, delicate style that may be tasted in any of the different types of wine in the Estate cellar. This extensive Estate produces sweet white and red wines, dry white, red and rosé wines and semi-sweet white wines. Most of the dessert wines are made in the fortified style from the slightly fermented juice of the Estate's Hanepoot and ripest Chenin blanc grapes. Recently, Prop and his cellar master, Johann Louwrens, have extended the range of sweet wines by producing a Noble Late Harvest wine from Chenin blanc grapes infected with *Botrytis cinerea*. Unlike the sweet wines fortified with spirit, the juice of these grapes is fermented until the combination of residual sugar and newly produced alcohol provides a medium that is too rich for the yeast to continue to function. At this stage of moderate alcohol and high sugar content the wine is cleaned and bottled.

Dry white wines are made from Chenin blanc, Colombar, Cape Riesling, Clairette blanche, Fernão Pires and Sauvignon blanc. To enhance the

All of Bergsig's vineyards receive additional water in summer. Some vineyards have permanent irrigation, the others use overhead sprinklers.

delicacy produced by the vineyards, Prop and Johann try to pick the white grapes at mid-ripe stage and allow the skins to stay mixed with the juice for at least half an hour. Free-run and lightly pressed juice are fermented together. The chilled juice is inoculated with yeast and fermented dry at relatively cold temperatures in closed stainless steel tanks. To make semi-sweet wines Johann stops the fermentation of chosen Chenin blanc and Colombar musts by removing the yeast, using a centrifuge and filter, at the desired balance of sugar and acidity. These wines are kept at low temperatures until they are despatched to the wholesaler or bottled for sale on the Estate.

Cabernet Sauvignon and Pinotage vineyards are used to make dry rosé and red wines in the soft and gentle Bergsig style. To make red wine, he prefers to have grapes measuring over 21° Balling, and ferments the juice and skins together in open concrete tanks at around 20 °C, retaining as much of the delicate nature of the grape as possible. To concentrate on the colour of red musts with low sugar content, Johann removes a small quantity of pink juice soon after crushing and ferments this dry to make rosé. The skins are removed from the red must as soon as an acceptable degree of colour has been extracted with minimum skin tannin, and the clear liquid is fermented dry. The wine is allowed to have a malolactic fermentation and is placed in 4 000-litre Nevers oak casks for a year of wood maturation before bottling. The vineyards produce light and soft red wines with delicate flavour that have proved to be very popular in the major domestic wine market of Johannesburg. The Cabernet Sauvignon rosé shares the early-matured soft drinkability of the red wine.

Further sweet wines are provided by the red vineyards, and Tinta Barocca, Cabernet Sauvignon and Malvasia Rey grapes are encouraged to over-ripen for the production of port. The juice is fermented with the skins in the mixture until about one third of the original sugar has been removed,

Within the narrow neck of land between two converging mountain ranges, the vineyards are regularly swept by summer breezes.

when the process is arrested by the addition of brandy spirit. This fortified wine stays in a closed steel tank for a few months until a portion is bottled and the remainder is transported to the wine merchant's depot in nearby Goudini.

Wine is for sale on the Estate.

On the Estate's higher ground, beside granite outcrops of rock, Hutton-type soils with relatively high clay content provide smaller yields of higher-quality grapes.

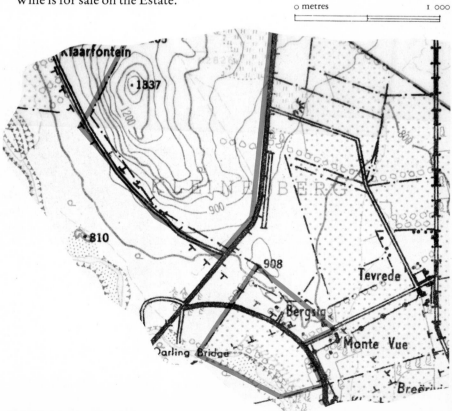

Lebensraum

The nation's leading Hanepoot producer, Phillip Deetlefs, has commenced a programme to make top-grade dry whites.

On Lebensraum winter growth between the vines is removed by hand.

The Lebensraum vineyards lie in the centre of the Goudini basin, at the foot of the Dutoitskloof Mountains.

After crossing the wide flat valley of the Breede River, the national highway approaches the Dutoitskloof Pass through a semi-circle of flat land criss-crossed with vineyards, known as Goudini. The Lebensraum wine Estate occupies a central position in the valley, alongside the Smalblaar River, and has the characteristic pebbly, deep-drained soils of this sheltered southern bank of the Breede. The mountains were crossed and this fertile area was settled from at least as early as 1730. Leendert Barents's farm 'Het Groot Eiland in de Slanghoek', of which Lebensraum was then a part, was undoubtedly approached from the Tulbagh Valley and Wolseley by the early Cape settlers and this accounts for the reference to the nearby Slanghoek Valley in the name. Later in the eighteenth century Groot Eiland was occupied by a Du Preez family. In 1807 it became the loan farm of Johannes du Preez, and was granted to him in 1831. His daughter Regina married Nicolaas Deetlefs in 1822, and on her father's death they were given a portion of Groot Eiland on which in that year Nicolaas planted the first vineyards. At the age of thirty-seven Deetlefs died of measles and in 1832 his widow built a substantial house, which is still standing on the farm today. In 1965 Phillipus Petrus Deetlefs inherited 160 hectares of Groot Eiland, together with the old house and cellar — a property that had been given the name Lebensraum by his father in 1947.

Lebensraum lies in a sheltered part of the western Cape, exposed to the moderate influence of the south-easterly, summer winds, and gets about 700 mm of rainfall each year, mostly in winter. The deep soil and the mounting summer temperatures rapidly deplete moisture and irrigation is essential for the growth of vineyards. Like most western Cape soils, Lebensraum's sandy loams require additional lime to raise their pH level and to encourage the vines to develop wide-ranging root systems.

One of the Cape's most famous styles of wine is made by blending the luscious grape juice of moderately sweet Muscat d'Alexandrie grapes with pure wine spirit. Traditionally, both the grape and the dessert wine are called Hanepoot. Unlike the Cape's other dessert wine style, Muscadel, where the grapes are ripened to the raisin stage, the most highly rated Hanepoot wines are made with grapes grown in the Goudini area and ripened to about 23° Balling before harvesting.

Lebensraum's Hanepoot vineyards have demonstrated a remarkable ability to concentrate the honeyed, Muscat flavour in their full-ripe grape, and Phillip Deetlefs has used these grapes to build a reputation as one of the nation's leading producers of Hanepoot wines.

Hanepoot is a prolific bearing vine, and in the early days of the wine industry was mainly used to make dry wine and wine for the distillation of brandy. The grapes were insufficiently sweet for the making of dessert wines or for the drying of raisins, two of the major grape industries of the early Cape. Hanepoot was most at home in the Goudini and Doornrivier areas beside the Dutoitskloof Mountains, and the grape growers in these areas were renowned for their muscat-flavoured raisins made from Hanepoot grapes. During the 1950s, sales of sweet dessert wines reached their peak and Phillip Deetlefs and some of his neighbours began to make dessert wine from the juice of Hanepoot grapes. They were encouraged to harvest the grapes at high sugar levels and to ferment this very sweet juice for about 4 ° Balling before the addition of pure wine spirit. The next three decades saw the peak and gradual decline of the dessert wine market, but the Goudini cellars, including Lebensraum, have continued as the foremost producers of Hanepoot and have established strong markets for the products of the area.

Phillip Deetlefs has modified the original technique used to make Hanepoot sweet wine. The Lebensraum Hanepoot grapes are picked at approximately the same sugar content as that required to make full-bodied white wines (22-23 ° Balling), are crushed, and the mixture of skins and grape juice is allowed to stand together overnight. The mash is put through a tank press the following morning and all the juice is extracted from the skins. This richly-flavoured juice is pumped into a closed concrete tank where the requisite amount of wine spirit is waiting. The alcohol content of the mixture that results is about 16,8 per cent, sufficiently high to prevent fermentation. The now fortified wine is left in the closed tank for about a month, then racked, fined and placed in a closed steel tank where it is allowed to mature for about one year before bottling. Lebensraum's Hanepoot is neither fermented nor aged in wood and has a delicate yet rich, spicy and honeyed flavour.

Though famed for sweet wines, the

Sheltered in a valley at the foot of the Dutoitskloof Mountains, Lebensraum has deep alluvial soils and moderate ripening temperatures.

Estate produces slightly more dry and semi-sweet wines than dessert wines and has vineyards of Chenin blanc, Sémillon, Clairette blanche and Colombard, in addition to Muscat d'Alexandrie. Most of Lebensraum's table wines are made in the light, early-drinkable style and the grapes of most vineyards are picked at early-ripe (17-19 ° Balling) stage. These grapes are crushed and their juice is inoculated with yeast and fermented dry at cool temperatures. Phillip also picks a portion of the Estate's Chenin blanc crop at full-ripe stage to produce a semi-sweet Stein wine.

He retains a small quantity of the sweet grape juice, unfermented, to add to the fermented wine, raising its sweetness to the required level. Dry Hanepoot wines are also made from the crop of vineyards that do not reach the required degree of sweetness to make the dessert wine.

Wine is for sale on the Estate.

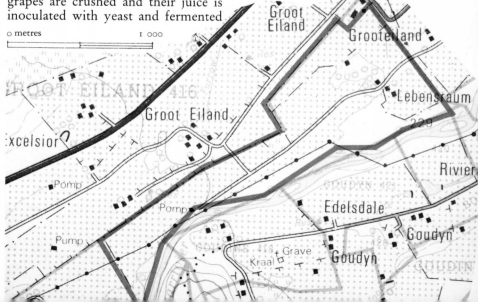

232 Worcester

Opstal

The sheer walls of the Dutoitskloof Mountains demonstrate how they reduce the hours of sunlight in Opstal's vineyards.

These vineyards on the eastern side of the river produce grapes for Hanepoot and Steen sweet wine.

Between Goudini and Wolseley, in a corner of the Dutoitskloof Mountains, a narrow valley is created by the Badsberg, a massive outcrop of granite. This north-south valley has a flat floor of alluvial soil, bisected by the Slanghoek River. The uniquely positioned Opstal Estate is situated in this valley, stretching from east to west, from high on the slopes of the Badsberg, across the valley floor and the river and up the sandstone cliffs of the Dutoitskloof Mountains.

The Slanghoek Valley has high rainfall, a very warm summer, but relatively short hours of direct sunlight. The sun rises behind the Badsberg, shines on the valley only during the warmest part of the day, then leaves the valley in shadow as it sets behind the Dutoitskloof. The southern part of the Slanghoek Valley gets twelve hours less direct sunlight per month than neighbouring farms in the open Breede River Valley. The south-easterly wind, normally a factor in lowering summer temperatures in the Cape, passes by the sheltered Slanghoek Valley and February temperatures are normally high.

"Our best wines are made in the cooler years," says Stanley Louw, Opstal's cellar master. "When the grapes are ripe, and we have a hot spell, the acidity drops quickly."

The whole valley was originally a rent farm used for cattle grazing. In the early 1700s Pieter du Toit, a Huguenot refugee, grazed cattle and made his base camp close to the spring on the Badsberg, near where the Opstal homestead stands today. The valley, known from Pieter du Toit's days as 'In de Slanghoek', was isolated by the Badsberg from the camps of other settlers and the farming conditions were hazardous. Before the advent of pumps and irrigation equipment, the combination of wet winters, deeply drained soils and hot, dry summers was the cause of many crop failures.

In 1796 the southern and larger portion of the Slanghoek farm was bought by four Rossouw brothers. A grandson of one of the brothers, Jan (known as Lang Jan), and six others became joint owners in 1849, and during their time the farm was subdivided. Lang Jan Rossouw had ten children, and his portion was divided into nine strips running from east to west across the valley, providing a farm for all but one of his children.

The Opstal Estate has been unchanged in size since that date and is owned by a family descended from Lang Jan Rossouw. Mrs. Attie Louw was born Ansie Everson, a great-great-grand-daughter of Lang Jan.

The advent of extensive irrigation, first from drainage furrows, and later using water pumped directly out of the river, changed the style of farming in the Slanghoek Valley. Orchards and vineyards were planted in the deep soils and watered through the long summer. The grapes were picked and left to dry to make raisins. Later the juice of a part of the grape crop was fermented and distilled to make brandy while the rest was fortified to make sweet wine.

Like many other Cape farmers during the 1920s Carel Everson, Mrs. Louw's grandfather, was often unable to sell his wine and had to run it out into the vineyards. Over-production, particularly in the Stellenbosch and Paarl areas, meant there was no market for wines from across the mountains. But when the market was receptive, he loaded wine in barrels on to an ox wagon, and carted his produce to Goudini Station. He normally

The high slopes of the Badsberg block out the early-morning sun.

The early heat and humidity of spring create ideal conditions for the outbreak of mildew diseases, and the vines must be protected.

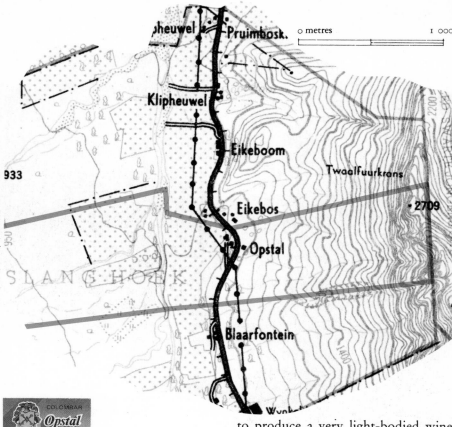

made about twelve leaguers of wine per year.

Today the Opstal Estate (of about 420 hectares, most of which is mountain) has about 90 hectares of vineyards on the lower slopes and valley floor. The major portion of the vineyards contains only three varieties: Chenin blanc, Colombard and Hanepoot. In addition, there are small vineyards of Clairette blanche and Cabernet Sauvignon. The difficulties of harvesting more than 40 hectares of Chenin blanc are lessened by the unusual topography and soil structure of the farm. The east-facing slopes of the valley have sandy loam soil, derived from the Dutoitskloof Mountains, that has difficulty in holding water for long in summer. Vines planted in this fertile soil, getting the morning sun, ripen earlier than their counterparts on the west-facing slopes, which are planted in soil of mixed decomposed granite and clay. The west-facing vineyards, benefiting from the warmer afternoon sun and the moisture retained by the clay, are able to reach higher concentrations of sugar. In general, Stanley Louw makes his fortified wines from the produce of the vineyards facing west and his table wines from the vineyards facing east.

These table wine vineyards rarely reach sugar contents of higher than 21° Balling, and produce relatively low-alcohol, gentle wines. The only red variety, Cabernet Sauvignon, is planted in the most elevated vineyard facing the setting sun, to enable this late-ripening variety to reach an optimum sugar content. These vines tend to produce a very light-bodied wine when fermented dry. Stanley prefers to allow the grapes to over-ripen and make a light-bodied port.

The white grapes chosen for fermentation are crushed and the juice is immediately drained off for settling. The juice is inoculated with yeast and then fermented at about 15 °C until all the sugar has been converted. A small quantity of grape juice is kept from the harvest to add to those wines that Stanley prefers to bottle as semi-sweet.

The Slanghoek area, like the Goudini area, is famous for sweet fortified Hanepoot wines. This variety, ripening late in the season, and planted on Opstal's clay soils, gains moderately high sugar content and produces a relatively light-bodied sweet wine, famous for its honeyed fragrance.

The 1978 harvest was chosen for the first bottling of Opstal's Estate wines and since then small quantities of dry, semi-sweet and sweet Hanepoot wines have been bottled for sale each year. The majority of Opstal's wines are still being sold to a Stellenbosch wine merchant.

Wine is for sale on the Estate.

Glossary

ACID Together with water and sugar, acid is one of three main constituents of the grape. As the grape ripens, the sugar content increases and the acid content decreases. In South Africa, where there are more than enough sunshine hours, the amount of acid in grapes usually determines the optimum time to pick the crop. Acid becomes one of the most important components of wine, determining the character as well as the balance of a specific wine. Wine with a low acidity tastes 'flat', and wine with too much acidity tastes hard and sharp. The acid is a natural preserving factor in the ageing of the wine. The fruity taste and crisp tang of a wine is due to acidity, usually between five and seven grammes per litre.

BALLING This term is used in measuring the sugar content or ripeness of the grapes, and takes its name from the surname of the inventor who devised the scale. The sugar content (in degrees Balling) is measured with a saccharometer, which floats in the liquid and is calibrated in degrees to show the density of the liquid. One degree Balling is equal to one per cent total extract, which consists of sugar, acid and other non-sugar extracts.

BOTTLE FERMENTATION A second sugar-alcohol fermentation induced artificially by the addition of sugar and yeast to wine inside a sealed bottle. On completion of the fermentation process, the yeast is removed from the wine by a gravitational clarification process, called *remuage* (shaking down), or by filtration. The former process is known as Méthode Champenoise, the latter as the transfer method.

BULK BEARERS Varieties which produce prolifically and are used mainly in the production of *vin ordinaire* and distilled spirit.

CAKE See cap.

CANES The vine is divided into roots, stem, arms and canes. The canes are new growth formed each year, and it is on canes that the grapes are formed.

CAP A cap or cake is an integral part of red wine fermentation. As carbon dioxide is formed in the fermenting must, the bubbles rise and push the skins towards the surface to form a cap on top of the liquid. This cap must be submerged frequently to increase the contact of skins and fermenting must.

CLONE A group of individual plants propagated asexually (vegetatively) from a single vine, in which all members of the group initially have characteristics identical to the parent vine.

CULTIVAR A variety cultivated for any purpose (i.e. not wild).

CUTTINGS Vine canes selected for grafting material.

DISTILLING WINE Wine not suitable for drinking, as well as low-grade wines produced from bulk bearers for the specific purpose of distillation into pure wine spirit.

DOWNY MILDEW (*Pernospora*) A fungus which can cause great damage to the vine, especially at the time of flowering. It can attack the flower and thereby destroy all of the crop, as diseased flowers do not form grapes. It can also damage the leaf structure and the green shoots. It spreads by infecting the bottom side of the leaf where it grows into the stomata and germinates. The fungus requires wet weather and must be combatted by using preventative sprays.

DRY WINE Wine resulting from a grape must in which the sugar has been allowed to ferment out completely. The use of the word 'dry' probably arises from the sensation on one's palate when drinking dry wine.

FERMENTATION The process in which yeast turns sugar into alcohol and carbon dioxide. Fermentation is part of nature's complete cycle in the life of the grape. The sugar contained in the grape originates from carbon dioxide and water, with carbon dioxide being absorbed by the vine leaf from the air, and water being provided by the vine roots from the ground. The green leaf uses energy from the sun to produce sugar through photosynthesis. In fermentation, the sugar is transformed by the yeast in the must to produce carbon dioxide and alcohol. The natural cycle, if not interrupted by man, would not end there. The alcohol would oxidise further and become acetic acid (vinegar) and finally carbon dioxide and water. During the process of breaking down the sugar molecules to its original constituents, energy is given off and the cycle is complete. Uncontrolled fermentation in South Africa's climate is rapid and not conducive to the production of quality wine. At high temperatures fermentation is fast, at low temperatures fermentation is slow. Cooling is used in all cellars to reduce the pace of fermentation, especially in the production of white wine. In the production of a white table wine, a must containing 22° Balling takes fourteen to twenty-one days to ferment out at ± 15° C. In the case of red wine, slightly higher temperatures, from 20° C to 25° C, are needed to have ideal conditions for the extraction of colour and robust flavour from the skins. Fermentation lasts from two to four days on skins, and after separation for about another seven days until fermented out dry.

FILTRATION Highly sophisticated filters have revolutionised the early cleaning up of wines. After fermentation, white wines in particular are fined and then filtered to separate them from old yeast cells, bacteria and other colloidal matter. This causes the wine to have a fresher, cleaner taste. Filtration is essential in larger wineries.

FINING A clearing process used soon after fermentation in which fining materials such as bentonite and gelatine are added to the wine. These substances react with excess protein and excess tannins to speed up the normal process of sedimentation, settling quickly to the bottom of the tank. Owing to the speed of this method, white wines are able to be marketed much sooner with more of their original fruity varietal character. Red wines still require maturation after fining to develop their character further.

FORTIFIED WINES Wine to which brandy or wine spirit has been added to stop fermentation or to increase the alcoholic content to a level at which it will act as a preservative. Sherries, ports, muscadels and hanepoots are examples of Cape fortified wines. They are also described as dessert or sweet wines.

GRAFTING Buds of European *Vitis vinifera* vines are grafted on to wild rootstock, in particular those of phylloxera-resistant American vines, to avoid the effects of phylloxera and facilitate a healthy union of the two different plants, towards which both contribute to give optimum production. All South African wine cultivars are grown with grafted American rootstock.

HEAT SUMMATION The practice of evaluating climatic zones by adding together the average daily temperature figures for various regions during the seven-month growing period of the vine, and comparing them. See page 28.

HUSKS The skins, pips and fleshy parts of the grape, which is reject matter in the production of white wine and is the source of colour and much of the flavour of red wine.

LEAGUER An old Cape wine measure, the contents of a large wooden cask. Roughly equivalent to a modern tonne.

LEES Sediment containing inert yeast cells, fruit tissue from the grapes, and in the case of red wine, pips and minute skin particles. Lees are collected and given a severe pressing to extract every drop of liquid which is normally used for distilling wine.

LUG BOXES Small boxes, normally rectangular, used for carrying cut bunches of grapes. Formerly wooden, these boxes are now invariably made of plastic. With a capacity of either 20 or 40 kilograms, the boxes prevent unwanted crushing of grapes by the weight of the load carried in larger containers.

MALOLACTIC FERMENTATION A secondary fermentation, occurring sometimes concurrently with, and often after, the first. Not a sugar-alcohol process, it is called malolactic fermentation as it is caused by bacteria attacking the malic acid and turning it into much weaker lactic acid and carbon dioxide gas, thereby lowering the total acid content of the wine. It occurs often in red wines and seldom in white wines in Cape cellars.

MATURATION Immediately after fermentation, all wines are raw and unpalatable to the taste. This is due to excess tannin and the fact that the alcohol formed in fermentation has not yet formed a union with the rest of the components of the wine. The problem is more apparent in red wines, which are often matured by the producer in semi-permeable wood, which allows a very limited contact with air, thus slightly increasing the rate of maturation. Maturation, as its name implies, requires time and is influenced by temperature. In general, the lower the temperature, the more gradual the maturation. Because of the delicacy of white wines, slow maturation is essential in their development, and they should be stored at a temperature between 12° C and 18° C. Red wine is more robust and is best matured at a temperature of from 15° C to 18° C. Bottle maturation of both red and white wines gives a more gradual and subtle final development to the wine.

MICRO-CLIMATE The immediate environment in which each vine grows and has to ripen its grapes. The micro-climate affects the grapes, and therefore the wine coming from individual vineyards, in different ways, even on the one Estate.

MUSCAT-TYPE A family of grape varieties belonging to *Vitis vinifera*, each of which produces grapes with a very specific aroma and taste. (An example is Muscat d'Alexandrie–Hanepoot.)

MUST The term used to describe the fermenting juice derived from crushed and pressed grapes.

NOBLE Usually applied to any grape variety, vineyard or wine that shows inherent, lasting superiority when compared with other examples of its type.

OENOLOGY The science of making wine.

OXIDATION The general term used when oxygen chemically reacts with other elements or compounds. Oxidation in wine involves a slightly more complicated chain of organic reactions, in which inorganic and organic catalysts cause oxygen to react with various constituents in the wine, and can occur at any time in the wine-making process and even within the bottle, if not prevented. Like all chemical reactions, the speed of oxidation is dependent on temperature. For this reason, we should store our wines (especially whites) at a cool temperature, to cause slow and gentle maturation. Uncontrolled oxidation causes browning and a bitter taste to form in the wine and is the reason why an unfortified wine cannot remain in contact with air for an indefinite time. Some oxidation is inevitable and can be seen as a slight yellowing or browning of the natural colour of the wine, and when noticeable usually indicates age.

pH A scale from 0 to 14 denoting the degree of acidity or alkalinity of any substance. Readings below 7 demonstrate a degree of acidity; those above 7 indicate a degree of alkalinity. Vines tend to prefer a fairly neutral soil, whereas wines need to be positively acid to be able to mature beneficially.

PHYLLOXERA An insect of North American origin that attacks the leaves and the roots of the vine, which it kills if the vine (e.g. *Vitis vinifera*) is not resistant to it.

POMACE The skin, pips and other residue of grapes after the juice has been extracted. Occasionally used for the production of strongly flavoured brandies, such as *marc*.

PROPAGATION The production of selected material from a chosen vine or vines, and the multiplication thereof to form vineyards.

RACKING The process of drawing off clear liquid from a tank of juice, must or wine, in which sediment or lees is left in the bottom of the container and thus separated. The process is usually repeated several times, before filtration.

REBATE WINE Wine specially prepared for distillation into brandy under strict control in pot stills.

REMUAGE A process of clarification of wine after it has undergone a secondary fermentation in the bottle, which does not involve filtration. *Remuage* means shaking and bottles of sparkling wine must be shaken in a particular manner. The bottles are placed almost horizontally and are gradually shaken and moved to a vertical position (cap down) to place the lees next to the opening. Today, both the traditional hand method and machines are used.

ROOTSTOCK Wild vines of American origin, which are naturally resistant to phylloxera; vines or cuttings of these vines, on to which scion (*Vitis vinifera*) cuttings are grafted and which produce the roots of the grafted vine.

SEMI-SWEET WINE Wines containing a small amount of natural grape sugar. These may be produced by having the fermentation arrested by chilling the must and preventing the remaining yeast from functioning or by blending a small quantity of sweet reserve with a wine that has been fermented dry.

SETTLING After the grapes have been crushed and pressed, the grape juice can contain fruit particles, bacteria, dust and colloidal matter, which may be detrimental to the wine if allowed into the fermentation process. The temperature of the juice is lowered to ± 11° C so that spontaneous fermentation cannot occur and the time gained allows unwanted matter to settle down by gravity. The next morning the clear juice can be racked off and pumped into the fermentation tank. The remainder can be cleaned to make a second-grade wine or used for distilling wine.

SHY-BEARING Varieties which have the characteristics of bearing small crops. Obviously, from an economic point of view, only the shy-bearing varieties that produce high-quality grapes are cultivated.

SKINS See husks.

SPONTANEOUS FERMENTATION The most natural form of wine making, where the yeast that starts the fermentation is present on the grape skins and in the grape juice produced when the grapes are crushed. Most wine makers prefer to remove or eliminate the wild yeast and work with specially cultured yeasts, selected for a specific purpose. However, several Estates use spontaneous fermentation successfully in the production of quality wine.

STABILISING Necessary to prevent instability in wines that have to be transported, especially when they may be subject to fluctuations in temperature. Modern technique chills the wine down to near freezing point before bottling so as to precipitate any crystals out before the wine reaches the bottle. These crystals are potassium salts and calcium salts of tartaric acid and are harmless. Stabilised wine is less likely to 'go off' even after long-distance transportation.

STORAGE There are ideal storage temperatures for both red and white wine. White wines are best kept at temperatures between 12° C and 16° C and red wines between 15° C and 18° C. A constant temperature is recommended even if these low temperatures are unattainable, with fluctuations in temperature to be avoided. Bottles should be kept on their side so that their corks cannot dry out, thus allowing free air contact to the wine.

SUGAR Only natural grape sugar, which is 50 per cent fructose and 50 per cent glucose, can be used in the production of wine. The sugar content is usually measured in degrees Balling.

SULPHUR The ancient Romans were the first to use sulphur to preserve their wine by burning sulphur in their amphorae (clay jugs) before filling and sealing them. Modern technique still uses sulphur in the form of sulphur dioxide (SO_2) to prevent oxidation and keep wines in a reductive, fruity state. The correct use of SO_2 is one of the crucial arts of the vintner.

SWEET WINE A fortified wine in which alcohol distilled from wine is added to fermenting grape juice to stop fermentation and keep the required amount of sweetness, whether the wine is to be full sweet (over 20° Balling) or half sweet (10 to 20° Balling). The volume percentage of alcohol must be over 16,5 to enable the wine to be classified as fortified. Sherries, ports and dessert wines may contain as much as 17,5 per cent by volume alcohol.

SWEET RESERVE Unfermented or partially fermented grape juice used to slightly sweeten wine that has been fermented out dry.

TABLE GRAPES Specific varieties chosen for the good looks and pleasant taste of their grapes, which are usually unsuitable for making quality wine.

TABLE WINE The quality classification given to unfortified wines ideal to drink at table with food.

TANNIN A group of organic compounds characterised by their astringent taste, occurring in wine where they are derived from the skin and seeds of the grape. The presence of tannin is usually noticeable only in red wine.

VARIETY A member of a vine species, having characteristics of leaf and grape that are consistent.

VINTNER The organiser and co-ordinator in the production of wine, from the planting of the vine to the bottling and maturation of the product.

VITICULTURE The science of the cultivation of vines.

VITIS VINIFERA The species of vine cultivated in Europe and Asia since time immemorial, chosen because of the superior quality of its grapes.

VLEI The Afrikaans word for a low-lying, poorly drained field.

WINKLER CLASSIFICATION See page 28.

YEAST A unicellular fungus to which we are deeply indebted for producing wine by breaking down the sugar molecule and giving off alcohol and carbon dioxide. There are many different species of yeast, of which some have been selected for specific applications. In South Africa's hot, dry summers, the sherry and port-producing yeasts, under similar conditions to those in Spain and Portugal, survive in the vineyard and come into the cellar with the grapes to start spontaneous fermentation. Most South African wines are made at temperatures lower than the environment and so require yeast specially chosen to work best in cool conditions. These cultured yeast strains are usually imported from Geisenheim in Germany, and Épernay and Bordeaux in France.

Index

Aan-De Doorns 21
Aan-den-Weg 132
Abercorn, Duke of 80
Act of Parliament 17, 19
Algeria 28
Allesverloren 209-12
Alphen 179, 191
Alsace 24
Alto 122-5, 173, 174, 175, 178
America 10
Amerine, Professor 28
Andalusia 21
Anglo American Corporation 59
Annandale farm 126
Anreith, Anton 8, 31, 193
Appel, Ferdinand 54, 200
Argentine 14
Arnold, Kevin 136, 137
Ashton 116, 117
auction wine 64
Audacia 126-7
Austen, Jane 30
Australia 10, 11, 14, 23, 27, 28, 178
Avonduur 205
Babylons Toren 54, 55, 69, 80
Back, Beryl 66
Back, C. L. 54, 64
Back, Charles 64, 66
Back, Cyril 64, 66, 67
Back, Michael 54, 55
Back, Sydney 54, 55, 56, 64
Backsberg 54-6, 64
Badsberg 233
Balling 236
Banhoek 205
Barents, Leendert 231
Barlow, Peter 179
Barlow family 136, 179
Barnard, Lady Anne 156
Barry, Charlotte 178
Barry, Sir Jacobus 178
Bax, Governor Johann 156
Beaujolais 26
Beit, Alfred 59
Bellinchamp 72, 80
Bellingham 73, 80
Benade, Dan 154
Benede-Oranje 21
Berg China 72
Bergkelder 81, 94, 108, 123, 125, 133, 150, 152, 195, 219
Berg River 68, 73, 169
Bergsig 229-30
Bertrams Wines 33
Bestbier, Johan 138
Bestbier, Petrus Johannes 132, 138
Beukman, Dirk 90
Beyers, Elizabeth 161
Beyers, Jan Andries 84
Beyers, Jan David 135, 164
Beyers, Johan 193
Beyers family 84, 164
Beyerskloof 153
Blaauwklippen 128-31, 173, 203
Blaauwklip River 129
Blanckenberg family 156
blending 35, 113, 130, 158
Blomkolsfontein farm 64
Blouberg, Battle of 157
Boberg 21
Boesmansrivier 21
Bon Courage 90-1, 107
Bonfoi 132-3
Bonnievale 21, 92, 99, 101, 107, 111, 113
Bonte River 126, 174, 175
Bonte Rivier 173
Boom, Coenraad 161
Boonzaier, Graham 129, 130
Boplaas 45-7, 48, 49
Bordeaux 86
Bordeaux blend 87
Bordeaux climate 9, 28

Bordeaux district 25, 28
Bordeaux style 172
Bordeaux wines 220
Boschendal 57-61, 75, 76, 80
Bosdari, C. de, *Wines of the Cape* 32
Bosman, Daniel 205
Bosman, Idee 142
Bosman, Isaac 140, 141
Bosman, Jacobus 142
Bosman, Johannes 142
Bosman, Michael 141, 142
Bosman, Petrus Jnr. 148
Bosman, Pieter 140, 141, 142
Bosman brothers 161
Bosman family 138, 140
Bossendal 57
Bothma, Steven Jansz 169
Botrytis cinerea 25, 94, 137, 188, 229
Bottelary farm 138, 148
Bottelary Hills 132, 138, 140, 141, 148, 161, 162, 166, 167, 183, 197, 205, 206
Bottelary Road 138
Bouwer, George 194
Bouwer, Gerry 195
Bouwer family 136
Brackenfell 140
Brandenburg, Jacoba 166
Brandvlei Dam 90, 93, 101, 111, 112, 114, 117
brandy 46, 50, 51, 69, 116, 127, 136, 232, 233
brandy, KWV 16, 216
brandy, rebate 102, 159
Breede River 21, 90, 91, 92, 96, 99, 101, 102, 107, 109, 110, 111, 114, 117
Breede River Valley 21, 226
Briet, Susanne 68
Brink, Arend 177
British Colonial Government 32
Brunswick 140
Bruwer, André 90, 91
Bruwer, Ernst 106
Bruwer, Pieter 106
Bruwer, Willie 90
Bruwer family 106
Buffelsvlei 45, 48
Bukettraube 23
bulk bearers 236
Burger, Jacobus 107, 108
Burger, Johannes 108
Burger, Johnny 107, 108
burgher 62, 72, 129, 156
Burgher Dragoons 129
Burgoyne's 123, 178
Burgundy district 28
Burgundy trophy 55
Bushmen 141, 209
Bush vines 144, 148, 149
By-den-Weg 128, 169
Cabernet franc 26
Cabernet Sauvignon 26
California 11, 23, 24, 27, 28, 58
Calitzdorp 45, 48, 49
Calitzdorp Co-operative 47
Camau, Jean de 72
Campher, Lorenz 135, 164
Canada 150
Canitz, Annemarie 165
Canitz, Georg Paul 135, 164, 165
canning industry 57, 59, 116
canteen wines 55
cap 236
Cape Agulhas 28, 226
Cape Championship 80
Cape Colony 8, 68, 80, 155, 156, 178
Cape Peninsula 31
Cape Province 33
Cape Riesling 23
Cape Town 33, 40, 46, 90
Cape trophy 80
Cape Wine and Distillers Ltd. 11
Cape Wine Show 123, 126,

127, 145, 200, 201, 210, 223, 224
Carignan 26
Cedarberg 21
cellar: equipment 47, 57, 59, 65, 69, 76, 114, 116, 125, 184, 191;
  technique 51, 76, 78, 215, 218
centrifuge 15, 95, 113, 137, 185, 230
Ceres-Karoo 154
certification 20
Chablis 23
Champagne: process 61;
  (district in France) 8, 23, 27, 28, 61, 68, 69;
  (land in Paarl) 57;
  method 19, 61, 68, 71, 185
Chardonnay 23
cheese: chevin 67; feta 67;
  goat's milk 66, 67
Chenin blanc 23
Chile 28
Cienska, Countess 80
Cinsaut 26
climate (general) 15, 18, 28, 33, 35, 86, 91, 94, 144, 145
Cloete, Hendrik jnr. 8, 177
Cloete, Hendrik snr. 32, 177, 182
Cloete, Jacob 177
Cloete, Peter 32
Cloete family 178
Cloete widow 209
clone 236
Coastal Region 21
Coetzee, Dirk 191
Coetzee, Jan Boland 146, 203
Coetzer, J. J. 45
Coetzenburg 191
cold settling 15
Colombard, Colombar 24
Colyn, Gilbert 205, 206, 207
Colyne, Johannes 200
concrete tanks 147, 165
Constantia 8, 13, 21, 29 (map), 30-5, 156, 179, 191
Constantiaberg 34
co-operatives 11
Council of Seventeen 40
Courtrai 62
Cradock, Sir John 203
cultivars: 20, 23-7;
  test programmes 13;
  breeding 14
Currey, J. B. 178
Customs and Excise 46
Dal Josaphat 28, 205
Dalla Cia, Giorgio 155
De Beers Diamond Company 59
De Chavonnes, M. 214, 215
De Drie Sprong 135
Deetlefs, Nicolaas 231
Deetlefs, Phillip 231, 232
Deetlefs, Phillipus Petrus 231
De Groot, Simon 183
De Hoop 182, 183
De la Batte, Jeanne 128
De la Noy, Nicholas 57
Delheim 134-7, 184
Denois, Jean-Louis 61
Department of Agriculture 33, 193
Depression, Great 46, 114, 132
Des Prez, Hercules 62
Des Prez family 62
De Rust 50
De Villiers, Abraham 57, 58, 151, 153
De Villiers, Baron 178
De Villiers, Gertruida 58
De Villiers, Hendrik 58
De Villiers, Hermanus 73
De Villiers, Hugo 78
De Villiers, Isaac 69
De Villiers, Jacob 151, 153
De Villiers, Jacques 58, 75, 76

De Villiers, Jan 58, 69, 166
De Villiers, Johannes 80
De Villiers, J. W. S. 84
De Villiers, Marguerite 58
De Villiers, Paul 58, 76, 77, 78
De Villiers, Pierre 58, 69, 166
De Villiers family 57, 75, 76, 78, 80, 151, 158, 166
Devon Valley 160, 161
De Vos, Pieter 200
dew 43, 59, 144, 157
De Waal, Chris 192
De Waal, Danie 191, 192
De Waal, Kottie 191
De Waal, Pieter ('Boy') 191
De Waal family 191, 192
De Wet, Albertus 101, 102
De Wet, Danie 92, 93, 94, 96, 118
De Wet, Freddie 97, 98
De Wet, Hendrik 101
De Wet, Jacobus 101
De Wet, Jacobus Stephanus 97
De Wet, Johann, 92, 93, 94, 96
De Wet, Kowie 97
De Wet, Paul jnr. 116, 117, 118, 119
De Wet, Paul snr. (Paulie) 116, 117, 118, 119
De Wet, Stephen 97, 98
De Wet, Wouter 101, 102
De Wet, Wouter de Vos 101
De Wet, Wouter Johannes 101
De Wet family 94, 116, 117
De Wetshof 92-6
De Zoete Inval 62-3
Die Krans 46, 47, 48-9
Diemer, Abraham 37
Diemersdal 37-9, 40
disease control 15, 18, 60, 234
disease test programme 13
distilled brandy 16, 46, 48, 101, 136
distilled spirit 48, 136
Distillers Corporation 125
distilling licence 114
Doornkraal 50-1
Douglas 21
Douglas Green 11
downy mildew 127, 236
drainage system 62
Driesprong 84, 134, 135, 136
Dumas, Cornelis jnr. 143, 144
Dumas, Cornelis snr. 143, 144
Dumas, Petrus Johannes (Hansie) 144
Du Preez, Johannes 231
Du Preez family 220, 231
Durbanville 10, 21, 36 (map), 37-43
Dutch East India Company 30, 40, 45, 48, 54, 57, 58, 62, 72, 79, 80, 107, 128, 152, 156, 193
Dutch East Indies 30
Dutch Reformed Church, Robertson 90
Du Toit, Gawie 99, 100
Du Toit, Gerhard 122
Du Toit, Hempies 125
Du Toit, Jan 100
Du Toit, Piet 122, 123, 124, 125
Du Toit, Pieter 233
Du Toit, Thys 100
earthquake 217, 222
Eckstein, Jacob 177
Eerste River 122, 129, 155, 156, 167, 188, 200
Eilandia 21
Elbertsz, Hendrik 169, 173
Ellis, Neil 34, 207
Elsenburg 84, 100, 103, 110, 135, 168, 193, 216
Engelbrecht, Jannie 168, 176
Estate farms 11, 18, 21, 59, 184
Estate wines 18-19, 73, 185
Esterhuyzen, Christoffel 161
Estreux, Christoffel 161

Europe 10, 30, 59, 178;
  climate 28
Everson, Ansie 233
Everson Carel 233
Excelsior 97-8
export brandy 46, 48
export wine 10, 33, 41, 54, 55, 63, 64, 68, 76, 183
Fairbridge, Dorothea 191
Fairview 64-7
False Bay 31, 123, 143, 144, 157, 189, 197, 200, 201
farming: dairy 38, 78, 226;
  peach 80, 116;
  orchards 54, 57, 59, 61, 117, 129, 226, 233;
  lucerne 50, 117;
  fruit 54, 59, 154;
  grazing 38, 48, 54, 90, 114, 132, 138;
  mixed 38, 40, 161;
  implements 79;
  techniques 182;
  wheat 38, 132, 138
Faure 155, 156
Faure, Brand 200, 201
Faure, Jac 200, 201, 202
Faure, Johannes Gysbertus 200
Faure, John 200, 201
Faure family 41, 127, 200, 201
fermentation: techniques 41, 43, 50, 60, 61, 63, 71, 76, 77, 83, 87, 91, 95, 96, 102, 113, 115, 124, 125, 133, 137, 141, 142, 144, 147, 152, 154, 158, 160, 162, 163, 168, 172, 175, 181, 193, 204, 215, 224, 227, 229, 232, 234;
  malolactic 11, 39, 43, 56, 71, 144, 147, 158, 230, 237;
  methods 39, 41, 47, 60, 65, 83, 160, 236;
  heat extraction 211, 212;
  spontaneous 35, 43, 165, 237;
  cold 71, 90, 116, 118, 141, 168, 184, 192, 223;
  tanks 43, 63, 65, 71, 124, 125, 137, 141, 168, 180, 192, 212, 230;
  cool 65, 224, 235;
  problems 118;
  secondary 71;
  bottle 18, 61, 70, 71, 184, 217, 236
fertiliser: inorganic 39, 81, 100, 130;
  organic 38, 39, 172, 218
filtration 236
fining 223, 236
Finlayson, Dr. M. H. 161, 162
Finlayson, Peter 163, 227
Finlayson, Walter 128, 130, 161, 162, 163
First World War 48, 116
floods 62, 119
flowering 12
fortified wines, sweet 33, 46, 48, 50, 51, 54, 55, 61, 63, 65, 76, 77, 90, 98, 101, 108, 109, 113, 114, 115, 116, 117, 143, 156, 159, 181, 209, 229, 232, 233, 234, 236
France 23, 26, 27, 30, 79, 168
Franschhoek 21, 79
Franschhoek Mountain, 82
Fransmanskraal 167
Frater, Adrian 62, 63
Frater, Gerard 63
Frater, John Robert 63
Frater, Robert 63
French oak 87, 113, 122, 158
Gabbema, Abraham 169
gable system 70
Gamay 23, 26
Geisenheim 73, 170

Geisz system 215
General Smuts trophy 63
Georgian architecture 193
Germany 24, 25, 28, 128 (Omnen), 128 (Straslund), 153, 168, 193, 201
Gewürztraminer 24
Gilbey's 11, 99, 130, 162, 163
Goede Hoop 72, 138-9
Goedemoed 90, 109
Goedverwacht 99-100, 110
Goree 21
Goudini 21, 231, 232, 233
Goudmyn 92, 93, 94, 108, 109
Graaff-Reinet 156
grafting 71, 184, 236
grape juice 50, 60, 114, 115, 141, 231, 232, 234
Great Karoo 50, 51
Greef, Matthys 182
Groenewald, Clara 187
Groenewald, Johannes 187
Groenfontein 48
Groenekloof 21
Groenrivier farm 122
Grommet, George 173
Groot Constantia 8, 30-5, 177, 193
Groot Constantia Control Board 34
Groot Constantia Estate Act 34
Groot Constantia Estate (Pty.) Ltd. 33
Groot Drakenstein 79, 81, 83
Haarhof, Frans 80
Hahn, Dr. Theophilus 183
Hamilton Russell, Tim 226
Hamilton Russell Vineyards 226-7
Hanekom, Herman 59, 60
Hanekom, Jurgen 151
Hanepoot 24
Hartenberg 161, 162
harvesting: methods 39, 63, 71, 86, 91, 100, 193, 215, 219, 220, 221, 222, 227, 231, 232;
  programmes 41, 60, 87, 91, 94, 95, 144;
  procedure 55;
  mechanical 24, 219
Hárslevelü 14
Hatting, Hans Hendrik 72, 187
Hauf, Mrs. 161
Hauf family 161
Haupt, Johannes 69
Hazendal 140-2, 206
Hazenwinkel, Christoffel 140
heat summation 28, 226, 237
Heemraad 129, 156
Helderberg Mountains 122, 126, 127, 132, 155, 173, 174, 176, 203, 205
Hellmer, Otto 136, 137
Hemel-en-Aarde Valley 226
Henning, Beyers 101
Hermanus 226-7
Hermitage 27
Het Hartenberg 161
Hiebner, Petrus 205
Hoffman, Dirk Wouter 129
Hoffman, Johann Bernard 128, 129, 187
Hoffman family 128
Hofmeyr, Billy 86, 87
Hofmeyr, Jan 109
Hofmeyr, Jan Hendrik 173, 174
Hofmeyr, Ursula 86, 87
Hoheisen, Del 134
Hoheisen, Hans 134, 135, 136
Holland 30, 40, 57, 68, 79, 109, 150, 156, 191, 200
Hoop op Constantia 34
Hoopsrivier 21
Hopefield 138
horses 116, 117
Hottentots 73

Hottentots Holland Mountains 169
Hugo, Anna 166
Hugo, Mr. 64
Huguenots 8, 30, 40, 57, 62, 68, 72, 73, 75, 79, 80, 164, 167, 187, 200, 233
Hungary 24
Hüsing, Henning 40, 54, 155, 156, 200
Hutton soils 37, 59, 67, 127, 146, 180, 185
Illing, Harvey 193, 194, 195
irrigation: 50, 51, 56, 70, 74, 75, 82, 101, 106, 114, 118, 155, 189, 226, 229, 233;
  spray 62, 63, 74, 100, 118, 157, 201, 229;
  drip 14, 62, 91, 100, 106, 118;
  flood 14, 90, 100;
  test programme 14;
  dam 11;
  sprinkler 14;
  micro-spray 14, 81
Italian Riesling 26
Italy 25, 28
Jacobsdal 143-4
Janssens, General 157
Jansz, Arnout Tamboer 72, 187
Jonker, Lourens 111, 112, 113
Jonker, Nicklaas 112
Jonker family 112
Jonkershuis 69, 173
Joostenberg 140
Joostenbergvlakte 84
Jordaan, A. W. 218
Jordaan, Kobus 218, 219
Jordaan, Rennie 218, 219
Jordaan family 218
Joubert, Chris 188, 189, 190
Joubert, Gideon 73
Joubert, Jean 214, 215
Joubert, Niel 184, 188, 190
Joubert, Pierre 72
Joubert, Pieter 72, 73, 79, 80
Joubert family 132, 169
Joubert widow 132
Kafferkuil 203
Kamanassie River 51
Kampen 191
Kanonkop 54, 145-7, 153, 203
Karoo 154; see Great Karoo, Little Karoo
Karoo soil 90, 117, 229
Kasteelberg 209
Kerner 24
Keyser, Paulus 161
Kimberlite 110
Klaas Voogds River 91, 107
Klapmutskop 86, 136, 151
Klawer 174
Kleigat 84
Klein Drakenstein 109
Klein Karoo, see Little Karoo
Klipdrift 114
Knorhoek 84, 135
Knysna 150
Koelenhof 148, 182, 183
Kogmanskloof River 97, 99, 111, 114, 117, 118
Kogmanskloof soils 110
Kohler, Dr. C. W. H. 16
Koopmanskloof 148-9
Krammer, Josef 84, 85
Kreft, Gerhard 215
Kriekbult 146
Krige, Jannie 146
Krige, Johannes 191, 192
Krige, Mary 146
Krige, Willem 191
Krige family 138
Kromme Rhee River 182
Krone, N. C. 220, 221, 222, 223
Krone, Nicky 220, 221, 223
Krone family 222

Kuils River 34, 140
KWV 10, 11, 16-17, 38, 39, 41, 46, 54, 63, 64, 68, 69, 70, 71, 76, 77, 90, 101, 138, 200, 201, 215, 216;
  wine appreciation courses 55, 69;
  wine experts 141
Laborie 68-71
labour problems 14, 189
La Brie 58, 68
La Concorde 68
Lady Anne Barnard's Bath 8
La Grange, Paul 132
La Motte 72-4, 80, 187
land grants 68
Landskroon 75-8, 86
Langeberg Mountains 91, 102, 107, 108
Langverwacht 112, 191, 205
La Provence 72, 73, 79, 80
La Rhone 218
La Rochelle 57, 58
Laszlo, Dr. Julius 94
Lategan, Johann 230
Lategan, Prop 229, 230
Lebensraum 231-2
Le Bonheur 150-2, 153
Le Chasseur 21, 101
Le Febre, Marie 80
Le Grand Chasseur 101-2
Leguat, François 69
Lekkerwyn, Ari 161
Le Long, Jean 57
Le Riche, Etienne 180, 181
Le Roux, Ernst 28
Le Roux, Gabriel 80
Le Roux, Gerrit 50, 51
Le Roux, Gert 50, 51
Le Roux, P.K. (Piet) 51
Le Roux, P.M. 50
Le Roux brothers 62
Libertas 40, 129, 187
Liebetrau, Johan 173
Liesbeek River 169
Lievland 153-4
Lindenhovious, Maria 156
Little Karoo, Klein Karoo 10, 21, 44 (map), 45-51, 48, 51
Livonia 153
loan farm 48, 73, 231
Loire Valley 25
London 46, 63, 123, 178
L'Ormarins 73, 74, 79-83
Loubser, Johannes 177
Loubser, Pieter 177
Lourmarin 79
Louw, Anna 173
Louw, Attie (Mrs.) 233
Louw, Beyers 38
Louw, Hendrik 69
Louw, Jacobus (Koos) 127
Louw, Jacobus (Kosie) 126, 127
Louw, Johannes 54, 69
Louw, Maria 69
Louw, Matthys 38
Louw, Stanley 233, 234
Louw, Tienie jnr. 38, 39
Louw, Tienie snr. 38
Louwrens, Johann 229
Lubbe, Aletta 161
Lubbe, Barend 161, 166
lug boxes 237
Madeira wine 80
mail order 46, 49, 114, 189
Malan, Daniel François 209, 210
Malan, Dr. D. F. 209
Malan, Fanie 210, 211, 212
Malan, Frans 183, 184, 186
Malan, Gerard 210
Malan, Hennie 122
Malan, Jacob 206
Malan, Johann 186
Malan, Liza 186
Malan, Manie 122, 123, 125
Malan, Pierre 72

Malan, Pieter 186
Malan, Stephanus 209, 210
Malan family 122, 209
Malmesbury 27
malolactic fermentation, see fermentation, malolactic
Marais, Charles 166, 167
Marais, Eksteen 114, 115
Marais, George 103
Marais, Hannetjie 103
Marais, Isaac 80
Marais, Johannes 80
Marais, Pierre 103, 104
Marais, Pieter 99
Marais family 79, 114, 167
Martin, John 55, 56
maturation: wood 39, 43, 47, 76, 83, 87, 95, 96, 113, 122, 137, 158, 165, 172, 176, 181, 186, 190, 217, 230, 237;
  wood and bottle 61, 119, 127, 185, 190;
  bottle 43, 46, 61, 63, 71, 96, 124, 176, 181, 216, 217, 218, 237;
  steel tank 56, 232;
  vats 125;
  technique 189, 237
Mauritius 30
McGregor 21
meat industry 156
mechanical: harvesting 14, 219;
  cultivation 125
Mediterranean climate 8
Meerland, Jan 40, 54, 68
Meerendal 37, 40-3, 68, 117
Meerlust 40, 155-8
Meiring, André 34
Melck, Martin 93, 129, 135, 193
Melck, Mrs. Martin 164
Merlot 26
Merriman, John X. 178
Méthode Champenoise 61, 68, 71, 185
Michell, Lewis 58
Michiels, Margharita 140
micro-climate 15, 236
Middelvlei 159-60, 166
mildew 234
Moller, Daniel 153
Momberg, Jan (of Middelvlei) 159, 160
Momberg, Jan (of Neethlingshof) 159, 167, 169
Momberg, Niels 159
Momberg, Tinnie 159
Mon Don 103-4
Moni, Roberto 165
Monica of the Coast 30
Monneaux 68
Montagne 130, 161-3
Mont Blois 105-6
Montpellier 214-17
Moorreesburg 100
Mossop and Frater 63
Mostert, Leon 47
Mowbray 169
Mulder vlei 41, 84, 193
Muller, Dr. H. 24
Muller, Inus 113
Muller, Janie 154
Müller-Thurgau 24
Muratie 16, 84, 135, 164-5
Muscat d'Alexandrie 24
Muscat Ottonel 25
must 63, 71, 77, 113, 124, 125, 149, 192, 237
Myburgh, Hannes 158
Myburgh, Johannes Albertus 156, 187
Myburgh, Nicolaas 155, 157, 158
Myburgh, Philip 156
Myburgh family 156, 157
Namaqualand 30
Napoleon Bonaparte 8, 30
Natal 150
Natalite 16
National monument 222
National Young Wine Show 55

Natte Vallei 151, 153
Naude, C. P. 84
Neethling, Johan 34
Neethling, Johannes 167
Neethlingshof 69, 126, 159, 161, 166-8
Nel, Aletta 107
Nel, Boets 49
Nel, Carel 47, 48, 49
Nel, Carel snr. 46, 48
Nel, Chris 47, 49
Nel, Danie 47, 49
Nel, Daniel jnr. 46, 48
Nel, Daniel snr. 46, 48
Nel, Guillaume 128
Nel, Louis 46
Nel family 46, 48
Nel's River 45, 46, 47, 48
Nevers Oak 61, 230
New Zealand 28
N G K Stellenbosch 129
Nicholson, Alfred 178, 179, 181
Nicholson, Reg 178, 179
Nicholson family 136
Niehaus, Dr. Charles 16, 17, 38, 41, 76, 77, 90, 138, 201
Nietvoorbij 12, 13
night-picking 220, 222
noble rot 19, 96, 113
Nooitgedacht 177, 182, 183
Normandie, Jean le Roux de 205
Nova Constantia 34
Nuy 21
Oenological and Viticultural Research Centre (OVRI) 12-15, 217;
  course 125
oenology 12;
  see also fermentation techniques;
  test programmes 14, 15
Olifantshoek 79
Olifants River 13, 21, 50, 51
O'Okiep 30
Oosterland 40, 68
Opstal 233-4
Orange Free State 129
Orange River 13, 156
ostrich feather industry 45, 46, 48, 49, 50, 51, 90, 114
Oude Meester 11, 17
Oude Weltevreden 151
Oudtshoorn 10, 109;
  experimental farm 13
Overberg 21
Overgaauw 128, 132, 169-72
oxidation 116, 215, 221, 222, 237
Paardenberg area 72
Paarl 21, 30, 38, 41, 52-3 (map), 54-87
Paarl Mountain 64, 66, 75, 169
Paarl Show 78
Palomino 25
Papegaaiberg 159
Parliament 51, 146
Pasman, Roelf 177
Pasman, Sibella 182
Pasman family 182
Perignon, Dom 68
Perold, Dr. 41, 215
Perold, Professor A. I. 16, 164, 165, 172, 201
Perold system 70
pH 237
phylloxera 10, 43, 178, 183, 221, 237
Picardie 68, 69
Piketberg 21
Pinotage 27
Pinot gris 25
Pinot noir 27
pine forest 136
planting: programmes 136, 144;
  techniques 82
Polkadraai 169, 191
pomace 237

Index 239

Pomerania 135
Ponty, C. E. 54
port and port-type wines 41, 51, 63, 64, 77, 78, 116, 130, 137, 138, 165, 170, 172, 210, 212, 230, 234
Portugal 28
Potgieter family 109
Pouilly-Fumé 25
press: 111;
   basket 87, 90
Prins, Ben 165
Provence 72, 79
pruning techniques 18, 74, 82, 91, 125, 138, 149, 211, 227
quitrent grants 54, 151, 159
quota 17
racehorses 116, 117
racking 237
rainfall in wine areas 10, 43, 49, 50, 56, 74, 81, 86, 116, 143, 153, 157, 180, 210, 221, 229, 231
raisins 46, 48
Rand Selection Corporation 59
Ratz, Helmut 84
rebate wine 90, 102, 116, 159, 237
Rembrandt 11, 17
remuage 71, 237
Retief, Hennie 109, 110
Retief, Nico 110
Retief, Nicolas 109
Retief, Wynand 109, 110
Reyne, Susanne 72
Rheinpfaltz 24, 136
Rhine Riesling 25
Rhodes, Cecil John 58, 129, 183
Rhodes Fruit Farms 57, 59
Rhone 127
Riebeekberg 21
Rietvallei 102, 107-8
Riverside 21, 112
Robbertsz, Pieter 177
Robertson: 114, 88-9 (map), 90-119;
   district 21, 108, 110;
   experimental farm 13
Roi, Jean 79, 80
Roodeberg 17
rootstocks 10, 14, 15, 43, 100, 152, 178, 184, 221, 226, 237
rosé 230
Rosendal 140
Rossouw, Jan 233
Rossouw, Pieter 92, 109
Rossouw brothers 233
Roux, Kobie 197, 198
Roux, Kosie 198, 199
Roux, Paul 179
Roux family 127
Rülander 25
Rupert, Anthony 73, 74, 81, 82, 83
Rupert, Dr. Anton 73, 79, 81
Rupert family 80
Rustenberg 177
Rust-en-Vrede 173-6
St. Emilion 25, 84
St. Helena 8, 30
Sainte Maure 67
Saldanha Bay 138
Salt River 38
Sanddrift Canal 99
Sandveld 132
Sauer, Paul 145, 146, 203
Sauer, Senator J. H. 145, 193
Sauer family 145, 193
Sauvignon blanc 25
Saxenburg 206
Schelde 62
Scherpenheuvel 21
Schoemanshoek 109
Scholms, Rudolf 205, 206, 207
Schoongezicht (Paarl) 76
Schoongezicht/Rustenberg 177-81
Schoongezicht (Stellenbosch) 135, 136, 153
scion 14

sea breezes 116, 117, 123, 144, 155, 157
Second World War 41, 115, 178
Sémillon 25
Sense and Sensibility, by Jane Austen 30
Serra, Alec 42
settling 237
sherry 25, 59, 76, 77, 78, 113
sherry, KWV 16, 38, 90, 201
Shiraz 27
Simondium 69
Simonsberg 64, 86, 135, 137, 151, 153, 164, 166, 177, 180, 183, 194, 197
Simonsberg (Stellenbosch) 21
Simonsig 182-6
Sint-Jans 62
Slanghoek 21
slaves 69, 73
Smit, Stewie 148, 149
Smit, Wynand 148
Smit family 148
Sneewind, Hendrik 37, 40
snow 117
Sonder Naam 197
soil: test programmes 13, 14, 122; loamy 81, 114; sandy 42, 62, 77, 78, 85, 126, 140, 143, 144, 149, 201, 231; sandy loam 47, 138, 152; Hutton 37, 38, 59, 67, 127, 130, 146, 153, 167, 171, 174, 180, 203, 206, 229; clay 42, 86, 148, 150, 151, 152, 227; Clovelly 174, 206; granite 66, 67, 68, 77, 123, 153, 161, 227; Karoo 47, 90, 91; limey 91, 98, 117; various 51, 56, 57, 59, 66, 70, 74, 98, 113, 114, 115, 126, 162, 163, 191, 227, 229, 234; types 59, 75, 160, 229; acid 85, 140; Breede River 91; stony 129; calcareous 96; Table Mountain sandstone 31; shale 161; pebbly 138; alluvial 11, 47, 51, 60, 62, 73, 90, 91, 96, 114, 117, 126, 129, 130, 201, 218, 232, 233; chalky 48; red 11, 38, 40, 41, 42, 43; saline 99; gravel 161, 210; preparation 38, 50, 51, 130, 227; fertile 50, 68, 114; correction 74, 146; deep-drained 63, 70, 74, 231; improvement 170
Sollier, Gilles 73
Somerset West 156
South African Breweries 17
South African Government 33
South African National Convention 178
South African wine industry 8-11, 33, 131
South West Africa 164
Souzão 27
Soviet Union 153
Spain 25, 28, 201
Sperling, Michael (Spatz) 134, 136, 184
Speyer 72
Spier 72, 129, 187-90, 192
stabilising: 238; tank 60
Starke, J. C. F. (Koosie) 41, 42, 43
Starke, William 40, 41
Steen 23
Steenkamp, Hans 154
Stellenbosch 13, 21, 30, 41, 43, 120-1 (map), 122-207
Stellenbosch Farmers' Wineries 11, 17
Stellenbosch Festival 128
Stellenbosch Kloof 132, 169, 191
Stellenbosch Mountain 130, 176, 203
Stellenbosch, University of 49, 51, 55, 101, 115, 118, 122, 123, 125, 144, 146, 164, 179, 200, 201, 210, 215

Stellenbosch Wine Route 184, 190, 198
Stellenbosch Young Wine Show 146
Stofberg, Koos 34
Stompdrift Dam 51
stone pines 168
storage 238
Storm, Jan David jnr. 101
sugar 71, 116, 238
sulphur 71, 80, 215, 216, 238
sultana grapes 46, 48
sunlight, hours of 153, 180, 233
'Superior' certification 20, 21, 113, 118, 119, 162, 189, 216, 217
Swakopmund 164
Swartberg 45, 46, 48
Swartland 21, 27, 208 (map), 209-12
Swaziland 150
sweet wine, see fortified wine
Swellendam 21
Table Bay 40, 68, 79, 80, 128, 191
Table Mountain 51
Table Mountain range 31, 32
Table Mountain sandstone 31
Taillefert, Isaac 68, 69, 70
Taillefert family 62, 69
tameletjies 47
tannin 238
Tas, Adam 40, 166, 200
tax, district 79
terracing 56, 68, 70, 81, 82, 221
Theron, Professor C. J. 123, 197, 210, 220
Theron, D. de Wet 214, 215, 216
Theron, Jan 216, 217
Theron, Hendrik 216, 217
Theron, Mynhardt 59
Theron family 214, 220, 221
Theron of La Rhone 220
Theuniskraal 123, 218-19
Thibault, L. M. 31, 193
Thierry, Château 68
tin mining 206
Tinta Barocca 27
Transvaal 129
trellising 70
Truter, Beyers 146, 147
Tulbagh 21, 123, 213 (map), 214-24
Twee Jongegezellen 123, 178, 218, 220-4
Tygerberg farm 40
Tygerberg Hills 37, 40, 41
Ugni blanc 251
Uiterwyk 128, 132, 191-2
Uitkyk 84, 128, 135, 136, 145, 193-6
Union of South Africa 178
Union Wine 11
United States 10, 14, 23, 28
Upington experimental farm 13
Vaalharts experimental farm 13
Van As family 140
Van Bochem, Jacob 161
Van Brakel, Adrian 173
Van Ceylon, Aron 161
Van der Byl, Andries 188
Van der Byl family 54, 188
Van der Helden, Jacobus 54, 156
Van der Merwe, Sibella 177
Van der Merwe, Sophia 177
Van der Merwe, Gert 154
Van der Stel, Simon 8, 30, 31, 57, 58, 72, 79, 80, 128, 169, 173, 182, 187, 191, 200
Van der Stel, Willem Adriaan 8, 40, 54, 62, 132, 135, 156, 164, 166, 177, 200, 203, 209
Van der Vyver, Gideon 108
Van der Vyver, Ria 108
Van der Wereld, Anna 172
Van der Wereld, Willem 173
Van der Westhuizen, Christoff 132, 133

Van der Westhuizen, George 132
Van der Westhuizen, Gys 167, 168
Van der Westhuizen, Johannes 132
Van der Westhuizen, Schalk 167, 168
Van der Westhuizen family 132
Van der Westhuyse, J. 37
Van Dijk, Hendrik 80
Van Loveren 109-10
Van Loveren, Christina 109
Van Niekerk family 193
Van Rheede, Jan 132
Van Rheenen, Joost Hendriks 166
Van Riebeeck, Jan 8, 10
Van Rooyen, Ronnie 59
Van Velden, Abraham Julius 169
Van Velden, Braam 171, 172
Van Velden, David 169, 171, 172
Van Wesel, Barend Pieterz 128
Van Wyk's River 62, 67
Van Ysel, Hendrina 187
Van Zyl, Guillaume 109
Van Zyl, Jacobus 90
Van Zyl, Jean 109
varieties 23, 35, 38, 43
Vera Cruz farm 134, 135, 136, 137
Verdun 197-9
Vergenoegd 200-2
Vermeulen, Japie 155
Viljoen, Wynand 165
Villiera 84-5
Vincent, Lady 80
vineyard test programme 15
Vinkrivier 21
vin ordinaire 24
vintage 20
Visser, Gerrit 128
viticulture 51, 58, 146
Vitis vinifera 13, 26, 238
Vlottenheimer 162
Von Arnim, Achim 57, 59, 60, 61
Von Carlowitz, George 135, 136, 193, 194, 195
Von Carlowitz, Hans 194
Von Carlowitz, Hilda 194
Von Stiernhjelm, Baron and Baroness Hendrika 153-4
Voogt, Claas 107
Vredendal 10, 74
Vredendal experimental farm 13
Vriesenhof 203-4
Waddinxveen 40
Walker Bay 21, 225 (map), 226-7
Watergang 148
Weisser Riesling 25
Welgeluk 46, 47, 49
Welgemeend 86-7
Welgevonden 138
Welmoed 156
Welsch Riesling 26
Weltevrede 111-13, 151, 161
Weltevreden 76, 151
Wemmershoek Dam 63
Werner, Karl 215, 216
Wessels, Hendrik 112
Western Province 31
wheat 79
wholesaling 11, 55
winds 43
Wine and Spirit Board 18, 19, 20
wine grapes 22
wine-making (oenology) test programme 13, 14
wine merchants (wholesalers) 11
wine seal 20-1, 23
'Wines of Origin' 18, 19, 21, 182
Wines of the Cape by C. de Bosdari 32

Winkler classification 28, 226
Winkler, Professor 28
Wolseley 73
Wolvendrift 99
Wolwedans 166, 167
Wonderfontein 114-15
Woodhead, Michael 150, 151, 152
Worcester 21, 46, 90, 101, 114, 228 (map), 229-34
Worcester fault 229
Worcester climate 9
yeast: 43, 60, 71, 76, 113, 165, 186, 200, 223, 229, 234, 238; flor 77, 141, 201
yoghurt 67
Yugoslavian oak 158
Zandvliet 97, 116-19
Zevenfontein 205
Zevenwacht 34, 205-7
Zinfandel 27
Zion 58

240   Index